Reasoning with Probabilistic and Deterministic Graphical Models: Exact Algorithms

Synthesis Lectures on Artificial Intelligence and Machine Learning

Reasoning with Probabilistic and Deterministic Graphical Models: Exact Algorithms
Rina Dechter
2013

A Concise Introduction to Models and Methods for Automated Planning
Hector Geffner and Blai Bonet
2013

Essential Principles for Autonomous Robotics
Henry Hexmoor
2013

Case-Based Reasoning: A Concise Introduction
Beatriz López
2013

Answer Set Solving in Practice
Martin Gebser, Roland Kaminski, Benjamin Kaufmann, and Torsten Schaub
2012

Planning with Markov Decision Processes: An AI Perspective
Mausam and Andrey Kolobov
2012

Active Learning
Burr Settles
2012

Computational Aspects of Cooperative Game Theory
Georgios Chalkiadakis, Edith Elkind, and Michael Wooldridge
2011

Representations and Techniques for 3D Object Recognition and Scene Interpretation
Derek Hoiem and Silvio Savarese
2011

A Short Introduction to Preferences: Between Artificial Intelligence and Social Choice
Francesca Rossi, Kristen Brent Venable, and Toby Walsh
2011

Human Computation
Edith Law and Luis von Ahn
2011

Trading Agents
Michael P. Wellman
2011

Visual Object Recognition
Kristen Grauman and Bastian Leibe
2011

Learning with Support Vector Machines
Colin Campbell and Yiming Ying
2011

Algorithms for Reinforcement Learning
Csaba Szepesvári
2010

Data Integration: The Relational Logic Approach
Michael Genesereth
2010

Markov Logic: An Interface Layer for Artificial Intelligence
Pedro Domingos and Daniel Lowd
2009

Introduction to Semi-Supervised Learning
XiaojinZhu and Andrew B.Goldberg
2009

Action Programming Languages
Michael Thielscher
2008

Representation Discovery using Harmonic Analysis
Sridhar Mahadevan
2008

Essentials of Game Theory: A Concise Multidisciplinary Introduction
Kevin Leyton-Brown and Yoav Shoham
2008

A Concise Introduction to Multiagent Systems and Distributed Artificial Intelligence
Nikos Vlassis
2007

Intelligent Autonomous Robotics: A Robot Soccer Case Study
Peter Stone
2007

Reasoning with Probabilistic and Deterministic Graphical Models: Exact Algorithms

Rina Dechter

www.morganclaypool.com

ISBN: 9781627051972 paperback
ISBN: 9781627051989 ebook

DOI 10.2200/S00529ED1V01Y201308AIM023

A Publication in the Morgan & Claypool Publishers series
SYNTHESIS LECTURES ON ARTIFICIAL INTELLIGENCE AND MACHINE LEARNING

Lecture #23
Series Editors: Ronald J. Brachman, *Yahoo Research*
William W. Cohen, *Carnegie Mellon University*
Peter Stone, *University of Texas at Austin*
Series ISSN
Synthesis Lectures on Artificial Intelligence and Machine Learning
Print 1939-4608 Electronic 1939-4616

Reasoning with Probabilistic and Deterministic Graphical Models: Exact Algorithms

Rina Dechter
University of California, Irvine

SYNTHESIS LECTURES ON ARTIFICIAL INTELLIGENCE AND MACHINE LEARNING #23

ABSTRACT

Graphical models (e.g., Bayesian and constraint networks, influence diagrams, and Markov decision processes) have become a central paradigm for knowledge representation and reasoning in both artificial intelligence and computer science in general. These models are used to perform many reasoning tasks, such as scheduling, planning and learning, diagnosis and prediction, design, hardware and software verification, and bioinformatics. These problems can be stated as the formal tasks of constraint satisfaction and satisfiability, combinatorial optimization, and probabilistic inference. It is well known that the tasks are computationally hard, but research during the past three decades has yielded a variety of principles and techniques that significantly advanced the state of the art.

In this book we provide comprehensive coverage of the primary exact algorithms for reasoning with such models. The main feature exploited by the algorithms is the model's graph. We present inference-based, message-passing schemes (e.g., variable-elimination) and search-based, conditioning schemes (e.g., cycle-cutset conditioning and AND/OR search). Each class possesses distinguished characteristics and in particular has different time vs. space behavior. We emphasize the dependence of both schemes on few graph parameters such as the treewidth, cycle-cutset, and (the pseudo-tree) height. We believe the principles outlined here would serve well in moving forward to approximation and anytime-based schemes. The target audience of this book is researchers and students in the artificial intelligence and machine learning area, and beyond.

KEYWORDS

graphical models, Bayesian networks, constraint networks, Markov networks, induced-width, treewidth, cycle-cutset, loop-cutset, pseudo-tree, bucket-elimination, variable-elimination, AND/OR search, conditioning, reasoning, inference, knowledge representation

Contents

Preface

Graphical models, including constraint networks (hard and soft), Bayesian networks, Markov random fields, and influence diagrams, have become a central paradigm for knowledge representation and reasoning, and provide powerful tools for solving problems in a variety of application domains, including scheduling and planning, coding and information theory, signal and image processing, data mining, computational biology, and computer vision.

These models can be acquired from experts or learned from data. Once a model is available, we need to be able to make deductions and to extract various types of information. We refer to this as *reasoning* in analogy with the human process of thinking and reasoning. These reasoning problems can be stated as the formal tasks of constraint satisfaction and satisfiability, combinatorial optimization, and probabilistic inference. It is well known that these tasks are computationally hard, but research during the past three decades has yielded a variety of effective principles and led to impressive scalability of exact techniques.

In this book we provide a comprehensive coverage of the main exact algorithms for reasoning with such models. The primary feature exploited by the algorithms is the model's graph structure and they are therefore uniformly applicable across a broad range of models, where dependencies are expressed as constraints, cost functions or probabilistic relationships. We also provide a glimpse into properties of the dependencies themselves, known as context-specific independencies, when treating deterministic functions such as constraints. Clearly, exact algorithms must be complemented by approximations. Indeed, we see this book as the first phase of a broader book that would cover approximation algorithms as well. We believe, however, that in order to have effective approximations we have to start with the best exact algorithms.

The book is organized into seven chapters and a conclusion. Chapter 1 provides an introduction to the book and its contents. Chapter 2 introduces the reader to the formal definition of the general graphical model and then describes the most common models, including constraint networks and probabilistic networks, which are used throughout the book. We distinguish two classes of algorithms: inference-based, message-passing schemes (Chapters 3, 4, and 5) and search-based, conditioning schemes (Chapters 6 and 7). This division is useful because algorithms in each class possesses common and distinguished characteristics and in particular have different behavior with respect to the tradeoff between time and memory. Chapter 7 focuses on this tradeoff, introducing hybrids of search and inference schemes. We emphasize the dependence of both types on few graph parameters such as the treewidth, cycle-cutset, and (the pseudo-tree) height.

The book is based on research done in my lab over the past two decades. It is largely founded on work with my graduate and postdoctoral students including: Dan Frost, Irina Rish, Kalev Kask, David Larkin, Robert Mateescu, Radu Marinescu, Bozhena Bidyuk, Vibhav Gogate, Lars Otten, Natasha Flerova and William Lam and my postdoctoral students Javier Larrosa, and Emma Rollon. Most heavily it relies on the work of Kalev Kask (Chapter 5) and Robert Mateescu (Chapters 6 and 7). I wish to also thank my colleagues at UCI for providing a supportive environment in our AI and machine learning labs, and especially to Alex Ihler for our recent collaboration that has been particularly inspiring and fruitful.

I owe a great deal to members of my family that took an active role in some parts of this book. First, to my son Eyal who spent several months reading and providing editing, as well as very useful suggestions regarding the book's content and exposition. Thanks also go to my husband Avi on providing editorial comments on large parts of this book and to Anat Gafni for her useful comments on Chapter 1.

Rina Dechter
Los Angeles, December 2013

CHAPTER 1

Introduction

Over the last three decades, research in artificial intelligence has witnessed marked growth in the core disciplines of knowledge representation, learning and reasoning. This growth has been facilitated by a set of graph-based representations and reasoning algorithms known as *graphical models*.

The term "graphical models" describes a methodology for representing information, or knowledge, and for reasoning about that knowledge for the purpose of making decisions by an intelligent agent. What makes these models *graphical* is that the structure of the knowledge can be captured by a graph. The primary benefits of graph-based representation of knowledge are that it allows compact encoding of complex information and its efficient processing.

1.1 PROBABILISTIC VS. DETERMINISTIC MODELS

The concept of graphical models has mostly been associated exclusively with *probabilistic graphical models*. Such models are used in situations where there is uncertainty about the state of the world. The knowledge represented by these models concerns the joint probability distribution of a set of variables. An unstructured representation of such a distribution would be a list of all possible value combinations and their respective probabilities. This representation would require a huge amount of space even for a moderate number of variables. Furthermore, reasoning about the information, for example, calculating the probability that a specific variable will have a particular value given some evidence would be very inefficient. A Bayesian network is a graph-based and far more compact representation of a joint probability distribution (and, as such, a graphical model) where the information is encoded by relatively small number of conditional probability distributions as illustrated by the following example based on the early example by Lauritzen and Spiegelhalter [Lauritzen and Spiegelhalter, 1988].

This simple medical diagnosis problem focuses on two diseases: lung cancer and bronchitis. There is one symptom, dyspnoea (shortness of breath), that may be associated with the presence of either disease (or both) and there are test results from X-rays that may be related to either cancer, or smoking, or both. Whether or not the patient is a smoker also affects the likelihood of a patient having the diseases and symptoms. When a patient presents a particular combination of symptoms and X-ray results it is usually impossible to say with certainty whether he suffers from either disease, from both, or from neither; at best, we would like to be able to calculate the probability of each of these possibilities. Calculating these probabilities (as well as many others) requires the knowledge of the joint probability distribution of the five variables (Lung Cancer

(L), Bronchitis (B), Dyspnea (D), Test of X-ray (T), and smoker (S)), that is, the probability of each of their 64 value combinations when we assume a bi-valued formulation for each variable (e.g., X-ray tests are either positive (value 1) or negative (value 0).

Alternatively, the joint probability distribution can be represented more compactly by factoring the distribution into a small number of conditional probabilities. One possible factorization, for example, is given by

$$P(S, L, B, D, T) = P(S)P(L|S)P(B|S)P(D|L, B)P(T|L) .$$

This factorization corresponds to the directed graph in Figure 1.1 where each variable is represented by a node and there is an arrow connecting any two variables that have direct probabilistic (and may be causal) interactions between them (that is, participate in one of the conditional probabilities).

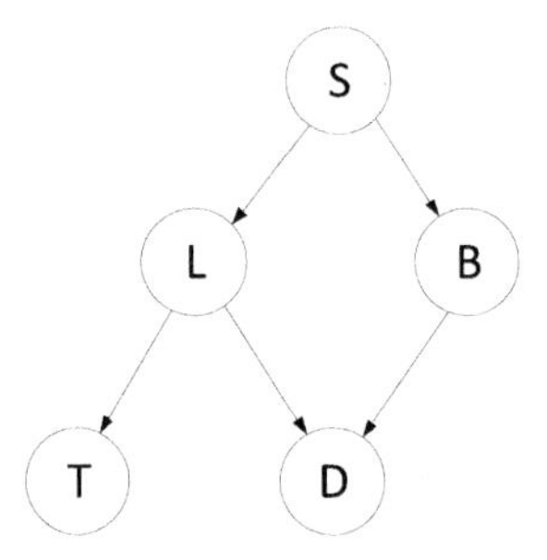

Figure 1.1: A simple medical diagnosis Bayesian network.

The graph articulates a more compact representation of the joint probability distribution, in that it represents a set of independencies that are true for the distribution. For example, it expresses that the variables lung cancer and bronchitis are conditionally independent on the variable smoking, that is, if smoking status is known then knowing that the patient has (or doesn't have) lung cancer has no bearing on the probability that he has bronchitis. However, if it is also known that shortness of breath is present, lung cancer and bronchitis are no longer independent; knowing that the person has lung cancer may explain away bronchitis and reduces the likelihood of dyspnea. Such dependencies and independencies are very helpful for reasoning about the knowledge.

While the term "graphical models" has mostly been used for probabilistic graphical models, the idea of using a graph-based structure for representing knowledge has been used with the same amount of success in situations that seemingly have nothing to do with probability distributions or uncertainty. One example is that of constraint satisfaction problems. Rather than the probability of every possible combination of values assigned to a set of variables, the knowledge encoded in a constraint satisfaction problem concerns their feasibility, that is, whether these value combination satisfy a set of constraints that are often defined on relatively small subsets of variables.

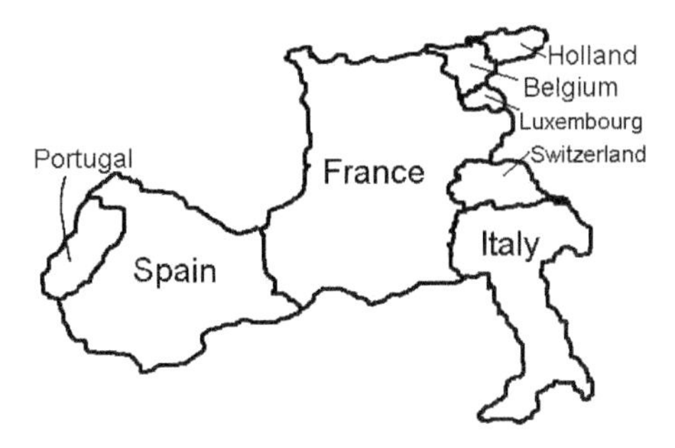

Figure 1.2: A map of eight neighboring countries.

The structure associated with these set of constraints is a constraint graph where each variable is represented by a node and two nodes are connected by an edge if they are bound by at least one constraint. A constraint satisfaction problem along with its constraint graph is often referred to as a constraint network and is illustrated by the following example.

Consider the map in Figure 1.2 showing eight neighboring countries and consider a set of three colors—red, blue, and yellow, for example. Each of the countries needs to be colored by one of the three colors so that no two countries that have a joint border have the same color. A basic question about this situation is to determine whether such a coloring scheme exists and, if so, to produce such a scheme. One way of answering these questions is to systematically generate all possible assignments of a color to a country and then test each one to determine whether it satisfies the constraint. Such an approach would be very inefficient because the number of different assignments could be huge. The structure of the problem, represented by its constraint graph in Figure 1.3, could be helpful in simplifying the task. In this graph each country is represented by a node and there is an edge connecting every pair of adjacent countries representing the constraint that prohibits that they be colored by the same color.

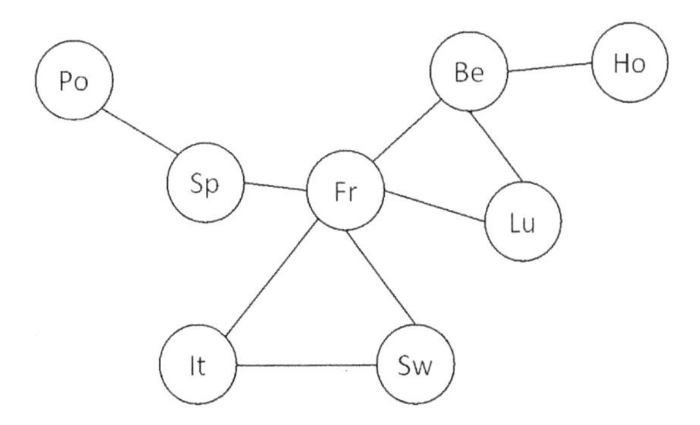

Figure 1.3: The map coloring constraint graph.

Just as in the Bayesian network graph, the constraint graph reveals some independencies in the map coloring problem. For example, it shows that if a color is selected for France the problem separates into three smaller problems (Portugal - Spain; Italy - Switzerland; and Belgium - Luxembourg - Holland) which could be solved independently of one another. This kind of information is extremely useful for expediting the solution of constraint satisfaction problems.

Whereas a Bayesian network is an example of a probabilistic graphical model, a constraint network is an example of a deterministic graphical model. The graphs associated with the two problems are also different: Bayesian networks use directed graphs, indicating that the information regarding relationship between two variables is not symmetrical while constraint graphs are undirected graphs. Despite these differences, the significance of the graph-based structure and the way it is used to facilitate reasoning about the knowledge are sufficiently similar to place both problems in a general class of graphical models. Many other problem domains have similar graph based structures and are, in the view of this book, graphical models. Examples include propositional logic, integer linear programming, Markov networks, and Influence Diagrams.

1.2 DIRECTED VS. UNDIRECTED MODELS

The examples in the previous section illustrate the two main classifications of graphical models. The first of these has to do with the kind information represented by the graph, primarily on whether the information is deterministic or probabilistic. Constraint networks are, for example, deterministic; an assignment of values to variables is either valid or not. Bayesian networks and Markov networks, on the other hand, represent probabilistic relationships; the nodes represent random variables and the graphical model as a whole encodes the joint probability distribution of those random variables. The distinction between these two categories of graphical models is not clear-cut, however. Cost networks, which represent preferences among assignments of values to variables are typically deterministic but they are similar to probabilistic networks as they are defined by real-valued functions just like probability functions.

The second classification of graphical models concerns how the information is encoded in the graph, primarily whether the edges in their graphical representation are directed or undirected. For example, Markov networks are probabilistic graphical models that have undirected edges while Bayesian networks are also probabilistic models but use a directed graph structure. Cost and constraint networks are primarily undirected yet some constraints are functional and can be associated with a directed model. For example, Boolean circuits encode functional constraints directed from inputs to outputs.

To make these classifications more concrete, consider a very simple example of a relationships between two variables. Suppose that we want to represent the logical relationship $A \vee B$ using a graphical model. We can do it by a constraint network of two variables and a single constraint (specifying that the relationship $A \vee B$ holds). The undirected graph representing this network is shown in Figure 1.4a. We can add a third variable, C, that will be "true" if an only if the relation $A \vee B$ is "true," that is, $C = A \vee B$. This model may be expressed as a constraint on all three variables, resulting in the complete graph shown in Figure 1.4b.

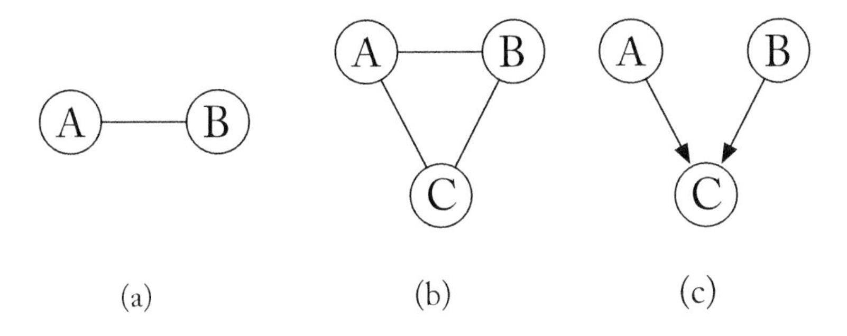

Figure 1.4: Undirected and directed deterministic relationships.

Now consider a probabilistic version of the above relationships, where the case of $C = A \vee B$ might employ a NOISY-OR relationship. A noisy-or function is the nondeterministic analog of the logical OR function and specifies that each input variable whose value is "1" produces an output of 1 with high probability $1 - \epsilon$ for some small ϵ. This can lead to the following encoding:

$$P(C = 1|A = 0, B = 0) = 0, \quad P(C = 1|A = 0, B = 1) = 1 - \epsilon_B,$$

$$P(C = 1|A = 1, B = 0) = 1 - \epsilon_A, \quad P(C = 1|A = 1, B = 1) = (1 - \epsilon_B)(1 - \epsilon_A) .$$

This relationship is directional, representing the conditional probability of C for any given inputs to A and B and can parameterize the directed graph representation as in Figure 1.4c. On the other hand, if we are interested in introducing some noise to an undirected relation $A \vee B$ we can do so by evaluating the strength of the OR relation in a way that fits our intuition or expertise, making sure that the resulting function is normalized. Namely, that the probabilities sum to 1. We could do the same for the ternary relation. These probabilistic functions are sometime called potentials or factors which frees them from the semantic coherency assumed when we talk about probabilities. Figure 1.5 shows a possible distribution of the noisy two- and three-variable OR relation, which is symmetrical.

From an algorithmic perspective, the division between directed and undirected graphical models is more salient and received considerable treatment in the literature [Pearl, 1988]. Deterministic information seems to be merely a limiting case of nondeterministic information where probability values are limited to 0 and 1. Alternatively, it can be perceived as the limiting cost in preference description moving from 2-valued preference (consistent and inconsistent) to multi-valued preference, also called *soft constraints*. Yet, this book will be focused primarily on methods that are indifferent to the directionality aspect of the models, and be more aware of the deterministic vs. non-deterministic distinction. The main examples used in this book will be constraint networks and Bayesian networks, since these are respective examples of both undirected and directed graphical models, and of Boolean vs. numerical graphical models.

A	B	$P(A \vee B)$
0	0	0
1	0	0.25
0	1	0.25
1	1	1/2

A	B	C	$P(A \vee B \vee C)$
0	0	0	0
1	0	0	1/15
0	1	0	1/15
0	0	1	1/15
1	1	0	2/15
1	0	1	2/15
0	1	1	2/15
1	1	1	6/15

Figure 1.5: Parameterizing directed and undirected probabilistic relations.

1.3 GENERAL GRAPHICAL MODELS

Graphical models include constraint networks [Dechter, 2003] defined by relations of allowed tuples, probabilistic networks [Pearl, 1988], defined by conditional probability tables over subsets of variables or by a set of potentials, cost networks defined by costs functions, and influence diagrams [Howard and Matheson, 1984] which include both probabilistic functions and cost functions (i.e., utilities) [Dechter, 2000]. Mixed networks is a graphical model that distinguish between probabilistic information and deterministic constraints. Each graphical model comes with its typical queries, such as finding a solution (over constraint networks), finding the most probable assignment, or updating the posterior probabilities given evidence, posed over probabilistic networks, or finding optimal solutions for cost networks.

The use of any model of knowledge (and graphical models are no exception) involves two largely independent activities, the construction of the model, and the extraction of useful information from the model. In the case of our medical diagnosis problem, for example, model construction involves the selection of the variables to be included, the structure of the Bayesian network, and the specification of the conditional probability distributions needed to specify the joint probability distribution. *Information extraction* involves answering queries about the effect of evidence on the probability of certain variables and about the best (most likely) explanation for such evidence. In the case of the map coloring problem, the model's structure is largely determined by the map to be colored. Information extraction involves answering queries like whether the map can be colored using a given set of colors, finding the minimum number of colors needed to color it, and, if a map cannot be colored by a given number of colors, finding the minimum number of constraint violations that have to be incurred in order to color the map.

The construction of the graphical model, including learning its structure and parameters from data or from experts, depends very much on the specific type of problem. For example, constructing a Bayesian network would be a very different process from constructing an integer linear programming optimization problem. In contrast, the process of answering queries over graphical models, in particular when taking advantage of their graph-based structure, is more universal and common in many respects across many types of problems. We call such activity as *reasoning* or *query processing*, that is, deriving new conclusions from facts or data represented explicitly in the models. The focus of this book is on the common reasoning methods that are used to extract information from given graphical models. Reasoning over probabilistic models is often referred to as *inference*. We, however, attribute a more narrow meaning to inference as discussed shortly.

Although the information extraction process for all the interesting questions posed over graphical models are computationally hard (i.e., NP-hard), and thus generally intractable, they invite effective algorithms for many graph structures as we show throughout the book. This includes answering optimization, constraint satisfaction, counting, and likelihood queries. The breadth of these queries render these algorithms applicable to a variety of fields including scheduling, planning, diagnosis, design, hardware and software testing, bio-informatics, and linkage analysis. Some learning tasks may be viewed as reasoning over a meta-level graphical model [Darwiche, 2009].

Our goal is to present a unifying treatment in a way that goes beyond a commitment to the particular types of knowledge expressed in the model. Previous books on graphical models focused either on probabilistic networks or on constraint networks. The current book is therefore broader in its unifying perspective. Yet it has restricted boundaries along the following dimensions. We address only graphical models over discrete variables (no continuous variables), cover only exact algorithms (a subsequent extension for approximation is forthcoming), and address only propositional graphical models (recent work on first-order graphical models is outside the scope of this book). In addition, we will *not* focus on exploiting the local structure of the functions, beyond our treatment of deterministic functions—a form of local structure. This is what is known as the context-specific information. Such techniques are orthogonal to graph-based principles and can, and should, be combined with them.

Finally, and as already noted, the book will not cover issues of modeling (by knowledge acquisition or learning from data) which are the two primary approaches for generating probabilistic graphical models. For this and more, we refer the readers to the books in the area. First and foremost is the classical book that introduced probabilistic graphical models [Pearl, 1988] and a sequence of books that followed amongst which are [Jensen, 2001; Neapolitan, 2000]. In particular, note the comprehensive two recent textbooks [Darwiche, 2009; Koller and Friedman, 2009]. For deterministic graphical models of Constraint networks see [Dechter, 2003].

1.4 INFERENCE AND SEARCH-BASED SCHEMES

As already noted, the focus of this book is on reasoning algorithms which exploit graph structures primarily and are thus applicable across all graphical models. These algorithms can be broadly classified as either *inference-based* or *search-based*, and each class will be discussed separately, as they have different characteristics. Inference-based algorithms perform a deductive step repeatedly while maintaining a single view of the model. Some example of inference-based algorithms we will focus on are resolution, variable-elimination and join-tree clustering. These algorithms are distinguished by generating new functions that augment the original model specification making it more explicit. By *inference* we also mean algorithms that reason by inducing equivalent model representations according to some set of inference rules. These are sometimes called *reparameterization schemes* because they generate an equivalent specification of the problem from which answers can be produced more easily. Inference algorithms are exponentially bounded in both time and space by a graph parameter called *treewidth*.

Search-based algorithms perform repeatedly a *conditioning step*, namely, fixing the value of a variable to a constant, and thus restrict the attention to a subproblem. This leads to a search over space of all subproblems. Search algorithms can be executed in linear space, a property that makes them particularly attractive. They can be shown to be exponentially bounded by graph-cutset parameters that depend on the memory level the algorithm would use. When search and inference algorithms are combined they enable improved performance by flexibly trading-off time and space. Search methods are more naturally poised to exploit the internal structure of the functions themselves, namely, their *local structure*. The thrust of advanced reasoning schemes is in combining inference and search yielding a spectrum of memory-sensitive algorithms applicable across many domains.

1.5 OVERVIEW OF THE BOOK

Chapter 2 introduces the reader to the graphical models framework and its most common specific models discussed throughout this book. This includes constraint networks, directed and undirected probabilistic networks, cost networks, and mixed networks. Influence diagram is an important graphical model combining probabilistic and cost information as well, which we dediced to not include here. Chapters 3, 4, and 5, focus on inference algorithms. Chapter 6 on search, while Chapter 7 concludes with hybrids of search and inference. Specifically, in the inference part, chapter 3 introduces the variable-elimination scheme called *bucket elimination (BE)* for constraint networks, and then Chapter 4 extends this scheme of bucket elimination to probabilistic networks, and to both optimization and likelihood queries. Chapter 5 shows how these variable elimination algorithms can be extended to message-passing scheme along tree-decompositions yielding the *bucket-tree elimination (BTE)*, *cluster-tree elimination (CTE)*, and the *join-tree or junction-tree* propagation schemes. Search is covered in Chapter 6 through the notion of AND/OR search spaces that facilitate exploiting problem decomposition within search schemes. Chapter 7 presents hybrids of search an inference whose main purpose is to design algorithms that can trade space for time and Chapter 8 provides some concluding remarks.

CHAPTER 2

What are Graphical Models

We will begin this chapter by introducing the general graphical model framework and continue with the most common types of graphical models, providing examples of each type: constraint networks [Dechter, 2003], Bayesian networks, Markov networks [Pearl, 1988], and cost networks. We also discuss a mix of probabilistic networks with constraints. Another more involved example which we will skip here is influence diagrams [Howard and Matheson, 1984].

2.1 GENERAL GRAPHICAL MODELS

Graphical models include constraint networks defined by relations of allowed tuples, probabilistic networks, defined by conditional probability tables over subsets of variables or by a set of potentials, cost networks defined by costs functions and mixed networks which is a graphical model that distinguish between probabilistic information and deterministic constraints. Each graphical model comes with its typical queries, such as finding a solution (over constraint networks), finding the most probable assignment or updating the posterior probabilities given evidence, posed over probabilistic networks, or finding optimal solutions for cost networks.

Simply put, a graphical model is a collection of *local* functions over subsets of variables that convey probabilistic, deterministic, or preferential information and whose structure is described by a graph. The graph captures independency or irrelevance information inherent in the model that can be useful for interpreting the data in the model and, most significantly, can be exploited by reasoning algorithms.

A graphical model is defined by a set of variables, their respective domains of values which we assume to be discrete, and by a set of functions. Each function is defined on a subset of the variables called its *scope*, which maps any assignment over its scope, an instantiation of the scopes' variables, to a real value. The set of local functions can be *combined* in a variety of ways (e.g., by sum or product) to generate a *global function* whose scope is the set of all variables. Therefore, a *combination* operator is a defining element in a graphical model. As noted, common combination operators are summation and multiplication, but we also have *AND* operator, for Boolean functions, or the relational *join*, when the functions are relations.

We denote variables or sets of variables by uppercase letters (e.g., X, Y, Z, S) and values of variables by lowercase letters (e.g., x, y, z, s). An assignment $(X_1 = x_1, ..., X_n = x_n)$ can be abbreviated as $\mathbf{x} = (x_1, ..., x_n)$. For a set of variables $\mathbf{S}$, $\mathbf{D}_S$ denotes the Cartesian product of the domains of variables in S. If $\mathbf{X} = \{X_1, ..., X_n\}$ and $\mathbf{S} \subseteq \mathbf{X}$, $\mathbf{x}_S$ denotes the restriction of $\mathbf{x} = (x_1, ..., x_n)$ to variables in $\mathbf{S}$ (also known as the projection of $\mathbf{x}$ over $\mathbf{S}$). We denote functions by letters f, g, h, etc., and the scope (set of arguments) of a function f by $scope(f)$. The projection of a tuple $\mathbf{x}$ on the scope of a function f, can also be denoted by $\mathbf{x}_{scope(f)}$ or, for brevity, by $\mathbf{x}_f$.

Definition 2.1 Elimination operators. Given a function h_S defined over a scope S, the functions ($\min_{\mathbf{X}} h$), ($\max_{\mathbf{X}} h$), and ($\sum_{\mathbf{X}} h$) where $\mathbf{X} \subseteq S$, are defined over $U = S - \mathbf{X}$ as follows: For every $U = u$, and denoting by $(\mathbf{u}, \mathbf{x})$ the extension of tuple $\mathbf{u}$ by the tuple $\mathbf{X} = \mathbf{x}$, $(\min_{\mathbf{X}} h)(\mathbf{u}) = \min_{\mathbf{x}} h(\mathbf{u}, \mathbf{x})$, $(\max_{\mathbf{X}} h)(\mathbf{u}) = \max_{\mathbf{x}} h(\mathbf{u}, \mathbf{x})$, and $(\sum_{\mathbf{X}} h)(\mathbf{u}) = \sum_{\mathbf{x}} h(\mathbf{u}, \mathbf{x})$. Given a set of functions $h_{S_1}, ..., h_{S_k}$ defined over the scopes $\mathcal{S} = \{S_1, ..., S_k\}$, the product function $\Pi_j h_{S_j}$ and the sum function $\sum_j h_{S_j}$ are defined over scope $U = \cup_j S_j$ such that for every $\mathbf{U} = \mathbf{u}$, $(\Pi_j h_{S_j})(\mathbf{u}) = \Pi_j h_{S_j}(\mathbf{u}_{S_j})$ and $(\sum_j h_{S_j})(\mathbf{u}) = \sum_j h_{S_j}(\mathbf{u}_{S_j})$. We will often denote h_{S_j} by h_j when the scope is clear from the context.

The formal definition of a graphical model is give next.

Definition 2.2 Graphical model. A *graphical model* $\mathcal{M}$ is a 4-tuple, $\mathcal{M} = \langle \mathbf{X}, \mathbf{D}, \mathbf{F}, \bigotimes \rangle$, where:

1. $\mathbf{X} = \{X_1, \ldots, X_n\}$ is a finite set of variables;
2. $\mathbf{D} = \{D_1, \ldots, D_n\}$ is the set of their respective finite domains of values;
3. $\mathbf{F} = \{f_1, \ldots, f_r\}$ is a set of positive real-valued discrete functions, defined over scopes of variables $\mathcal{S} = \{S_1, ..., S_r\}$, where $\mathbf{S}_i \subseteq \mathbf{X}$. They are called *local* functions.
4. $\bigotimes$ is a *combination* operator (e.g., $\bigotimes \in \{\prod, \sum, \bowtie\}$ (product, sum, join)). The combination operator can also be defined axiomatically as in [Shenoy, 1992], but for the sake of our discussion we can define it explicitly, by enumeration.

The graphical model represents a *global function* whose scope is $\mathbf{X}$ which is the combination of all its functions: $\bigotimes_{i=1}^{r} f_i$.

Note that the local functions define the graphical model and are given as input. The global function provides the meaning of the graphical model but it cannot be computed explicitly (e.g., in a tabular form) due to its exponential size. Yet all the interesting reasoning tasks (called also "problems" or "queries") are defined relative to the global function. For instance, we may seek an assignment on all the variables (sometime called configuration, or a solution) having the *maximum* global value. Alternatively, we can ask for the number of solutions to a constraint problem, defined by a *summation*. We can therefore define a variety of *reasoning queries* using an additional operator called *marginalization*. For example, if we have a function defined on two variables, $F(X, Y)$, a maximization query can be specified by applying the *max* operator written as $\max_{x,y} F(x, y)$ which returns a function with no arguments, namely, a constant, or we may seek the maximizing tuple $(x^*, y^*) = argmax_{x,y} F(x, y)$. Sometimes we are interested to get $Y(x) = argmax_y F(x, y)$.

Since the marginalization operator, which is max in the above examples, operates on a function of several variables and returns a function on their subset, it can be viewed as *eliminating* some variables from the function's scope to which it is applied. Because of that it is also called an *elimination* operator. Consider another example when we have a joint probability distribution on two variables $P(X, Y)$ and we want to compute the marginal probability $P(X) = \sum_y P(X, y)$. In this case, we use the *sum* marginalization operator to express our query. A formal definition of a reasoning task using the notion of a *marginalization* operator, is given next. We define *marginalization* by explicitly listing the specific operators we consider, but those can also be characterized axiomatically ([Bistarelli *et al.*, 1997; Kask and Dechter, 2005; Shenoy, 1992]).

Definition 2.3 A reasoning problem. A *reasoning problem* over a graphical model $\mathcal{M} = \langle \mathbf{X}, \mathbf{D}, \mathbf{F}, \bigotimes \rangle$ and given a subset of variables $\mathbf{Y} \subset \mathbf{X}$ is defined by a marginalization operator $\Downarrow_{\mathbf{Y}}$ explicitly as follows. $\Downarrow_{\mathbf{Y}} f_{\mathbf{S}} \in \{\max_{\mathbf{S}-\mathbf{Y}} f_{\mathbf{S}}, \min_{\mathbf{S}-\mathbf{Y}} f_{\mathbf{S}}, \pi_{\mathbf{Y}} f_{\mathbf{S}}, \sum_{\mathbf{S}-\mathbf{Y}} f_{\mathbf{S}}\}$ is a *marginalization* operator. The reasoning problem $\mathcal{P}\langle \mathcal{M}, \Downarrow_{\mathbf{Z}} \rangle$ for a scope $\mathbf{Z} \subseteq X$ is the task of computing the function $\mathcal{P}_{\mathcal{M}}(\mathbf{Z}) = \Downarrow_{\mathbf{Z}} \bigotimes_{i=1}^{r} f_i$, where r is the number of functions in F.

Many reasoning problems are defined by $\mathbf{Z} = \{\emptyset\}$. Note that in our definition $\pi_{\mathbf{Y}} f$ is the relational projection operator (to be defined shortly) and unlike the rest of the marginalization operators the convention is that it is specified by the scope of variables that are *not* eliminated.

2.2 THE GRAPHS OF GRAPHICAL MODELS

As we will see throughout the book, the structure of graphical models can be described by graphs that capture dependencies and independencies in the knowledge base. These graphs are useful because they convey information regarding the interaction between variables and allow efficient query processing.

2.2.1 BASIC DEFINITIONS

Although we already assumed familiarity with the notion of a graph, we take the opportunity to define it formally now.

Definition 2.4 Directed and undirected graphs. A *directed graph* is a pair $G = \{V, E\}$, where $V = \{X_1, \ldots, X_n\}$ is a set of vertices and $E = \{(X_i, X_j) | X_i, X_j \in V\}$ is the set of edges (arcs). If $(X_i, X_j) \in E$, we say that X_i *points to* X_j. The degree of a variable is the number of arcs incident to it. For each variable, X_i, $pa(X_i)$, or pa_i is the set of variables pointing to X_i in G, while the set of child vertices of X_i, denoted $ch(X_i)$, comprises the variables that X_i points to. The family of X_i, F_i, includes X_i and its parent variables. A directed graph is acyclic if it has no directed cycles. An *undirected graph* is defined similarly to a directed graph, but there is no directionality associated with the edges.

A graphical model can be represented by a *primal graph*. The absence of an arc between two nodes indicates that there is no direct function specified between the corresponding variables.

Definition 2.5 Primal graph. The *primal graph* of a graphical model is an undirected graph that has variables as its vertices and an edge connects any two variables that appear in the scope of the same function.

The primal graph (also called moral graph for Bayesian networks) is an effective way to capture the structure of the knowledge. In particular, graph separation is a sound way to capture conditional independencies relative to probability distributions over directed and undirected graphical models. In the context of probabilistic graphical models, primal graphs are also called i-maps (independence maps [Pearl, 1988]). In the context of relational databases [Maier, 1983], primal graphs capture the notion of embedded multi-valued dependencies (EMVDs).

All advanced algorithms for graphical models exploit their graphical structure. Besides the primal graph, other graph depictions include hyper-graphs, dual graphs, and factor graphs.

2.2.2 TYPES OF GRAPHS

The arcs of the primal graph do not provide a one to one correspondence with scopes. Hypergraphs and dual graphs are representations that provide such one-to-one correspondence.

Definition 2.6 Hypergraph. A hypergraph is a pair $\mathcal{H} = (V, S)$ where $V = \{v_1, ..., v_n\}$ is a set of nodes and $S = \{S_1, ..., S_l\}$, $S_i \subseteq V$, is a set of subsets of V called hyperedges.

A related representation that converts a hypergraph into a regular graph is the *dual graph*.

Definition 2.7 A dual graph. A hypergraph $\mathcal{H} = (V, S)$ can be mapped to a *dual graph* $\mathcal{H}^{dual} = (S, E)$ where the nodes of the dual graph are the hyperedges $S = \{S_1, ..., S_l\}$ in $\mathcal{H}$, and $(S_i, S_j) \in E$ iff $S_i \cap S_j \neq \emptyset$.

Definition 2.8 A primal graph of a hypergraph. A primal graph of a hypergraph $\mathcal{H} = (V, S)$ has V as its set of nodes, and any two nodes are connected if they appear in the same hyperedge.

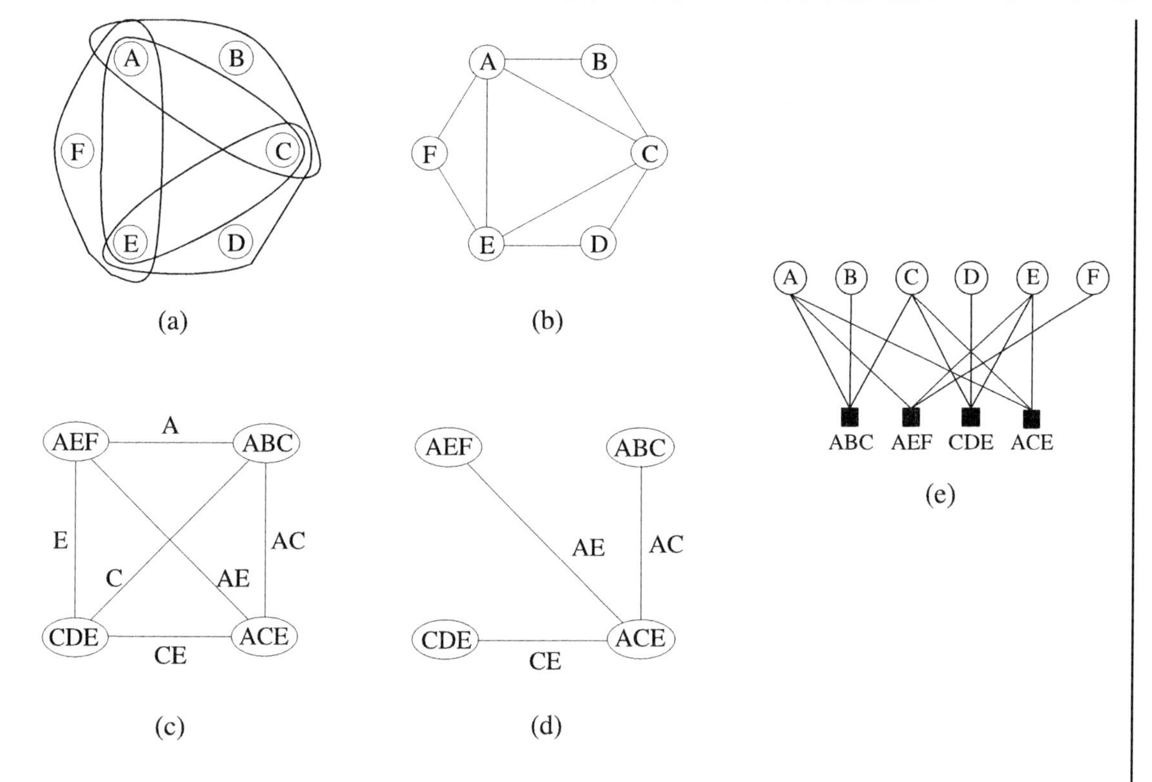

Figure 2.1: (a) Hyper; (b) primal; (c) dual; (d) join-tree of a graphical model having scopes ABC, AEF, CDE and ACE; and (e) the factor graph.

GRAPHICAL MODELS AND HYPERGRAPHS

Any graphical model $\mathcal{M} = \langle \mathbf{X}, \mathbf{D}, \mathbf{F}, \bigotimes \rangle, \mathbf{F} = \{f_{\mathbf{S}_1}, ..., f_{\mathbf{S}_t}\}$ can be associated with a hypergraph $\mathcal{H}_{\mathcal{M}} = (\mathbf{X}, H)$, where $\mathbf{X}$ is the set of nodes (variables), and H is the scopes of the functions in $\mathbf{F}$, namely $H = \{\mathbf{S}_1, ..., \mathbf{S}_l\}$. Therefore, the dual graph of (the hypergraph of) a graphical model associates a node with each function's scope and an arc with each two nodes corresponding to scopes sharing variables.

Example 2.9 Figure 2.1 depicts the *hypergraph* (a), the *primal graph* (b), and the *dual graph* (c) representations of a graphical model with variables A, B, C, D, E, F and with functions on the scopes (ABC), (AEF), (CDE), and (ACE). The specific functions are irrelevant to the current discussion; they can be arbitrary relations over domains of $\{0, 1\}$, such as $C = A \vee B$, $F = A \vee E$, CPTs or cost functions.

A factor graph is also a popular graphical depiction of a graphical model.

Definition 2.10 Factor graph. Given a graphical model and its hypergraph $H = (V, S)$ defined by the functions scopes, the factor graph has function nodes and variable nodes. Each scope is associated with a function node and it is connected to all the variable nodes appearing in the scope. Figure 2.1e depicts the factor graph of the hypergraph in part (a). The meaning of graph (d) will be described shortly.

We will now describe several types of graphical models and show how they fit the general definition.

2.3 CONSTRAINT NETWORKS

Constraint networks provide a framework for formulating real world problems as satisfying a set of constraints among variables, and they are the simplest and most computationally tractable of the graphical models we will be considering. Problems in scheduling, design, planning, and diagnosis are often encountered in real-world scenarios and can be effectively rendered as constraint networks problems.

Let's take scheduling as an example. Consider the problem of scheduling several tasks, where each takes a certain time and each have different options for starting time. Tasks can be executed simultaneously, subject to some precedence restriction between them due to certain resources that they need but cannot share. One approach to formulating such a scheduling problem is as a constraint satisfaction problem having a variable for each combination of resource and time slice (e.g., the conference room at 3 p.m. on Tuesday, for a class scheduling problem). The domain of each variable is the set of tasks that need to be scheduled, and assigning a task to a variable means that this task will begin at this resource at the specified time. In this model, various physical constraints can be described as constraints between variables (e.g., that a given task takes 3 h to complete or that another task can be completed at most once).

The *constraint satisfaction task* is to find a solution to the constraint problem, that is, an assignment of a value to each variable such that no constraint is violated. If no such assignment can be found, we conclude that the problem is inconsistent. Other queries include finding all the solutions and counting them or, if the problem is inconsistent, finding a solution that satisfies the maximum number of constraints.

Definition 2.11 Constraint network. A *constraint network (CN)* is a 4-tuple, $\mathcal{R} = \langle \mathbf{X}, \mathbf{D}, \mathbf{C}, \bowtie \rangle$, where $\mathbf{X}$ is a set of variables $\mathbf{X} = \{X_1, \ldots, X_n\}$, associated with a set of discrete-valued domains, $\mathbf{D} = \{D_1, \ldots, D_n\}$, and a set of constraints $\mathbf{C} = \{C_1, \ldots, C_r\}$. Each constraint C_i is a pair $(\mathbf{S}_i, R_i)$, where R_i is a relation $R_i \subseteq D_{\mathbf{S}_i}$ defined on scope $\mathbf{S}_i \subseteq \mathbf{X}$. The relation denotes all compatible tuples of $D_{\mathbf{S}_i}$ allowed by the constraint. The *join* operator $\bowtie$ is used to combine the constraints into a global relation. When it is clear that we discuss constraints we will refer to the problem as a triplet $\mathcal{R} = \langle \mathbf{X}, \mathbf{D}, \mathbf{C} \rangle$. A *solution* is an assignment of values to all the variables, denoted $\mathbf{x} = (x_1, \ldots, x_n)$, $x_i \in D_i$, such that $\forall\ C_i \in \mathbf{C}$, $\mathbf{x}_{\mathbf{S}_i} \in R_i$. The constraint network represents its set of solutions, $sol(\mathcal{R}) = \bowtie_i R_i$. Therefore, a *constraint network* is a graphical model

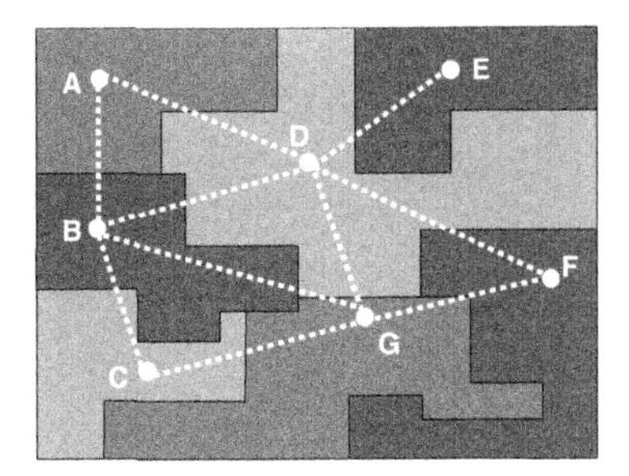

(a) Graph coloring problem

Figure 2.2: A constraint network example of a map coloring.

whose functions are relations and whose combination operator is the relational join ($\bigotimes = \bowtie$). The primal graph of a constraint network is called a *constraint graph*. It is an undirected graph in which each vertex corresponds to a variable and an edge connects any two vertices if the corresponding variables appear in the scope of the same constraint.

Example 2.12 The map coloring problem in Figure 2.2(a) can be modeled by a constraint network: given a map of regions and three colors {red, green, blue}, the problem is to color each region by one of the colors such that neighboring regions have different colors. Each region is a variable, and each has the domain {red, green, blue}. The set of constraints is the set of relations *"different"* between neighboring regions. Figure 2.2 overlays the corresponding constraint graph and one solution (A=red, B=blue, C=green, D=green, E=blue, F=blue, G=red) is given. The set of constraints are $A \neq B$, $A \neq D$, $B \neq D$, $B \neq C$, $B \neq G$,$D \neq G$, $D \neq F$,$G \neq F$, $D \neq E$.[1]

Next we define queries of constraint networks.

Definition 2.13 The queries, or constraint satisfaction problems. The primary query over a constraint network is deciding if it has a solution. Other relevant queries are enumerating or counting the solutions. Those queries over $\mathcal{R}$ can be expressed as $\mathcal{P} = \langle \mathcal{R}, \pi, \mathbf{Z} \rangle$, when marginalization is the relational projection operator π. That is, $\Downarrow_{\mathbf{Y}} = \pi_{\mathbf{Y}}$. For example, the task of enumerating all solutions is expressed by $\Downarrow_{\emptyset} \bigotimes_i f_i = \pi_{\emptyset}(\bowtie_i f_i)$. Another query is to find the minimal domains of variables. The minimal domain of a variable X is all its values that participate in any solution. Using relational operations, $MinDom(X_i) = \pi_{X_i}(\bowtie_j R_j)$.

Example 2.14 As noted earlier, constraint networks are particularly useful for expressing and solving scheduling problems. Consider the problem of scheduling five tasks (T1, T2, T3, T4,

[1]Example taken from [Dechter, 2003].

T5), each of which takes one hour to complete. The tasks may start at 1:00, 2:00, or 3:00. Tasks can be executed simultaneously subject to the restrictions that:

- T1 must start after T3;
- T3 must start before T4 and after T5;
- T2 cannot be executed at the same time as either T1 or T4; and
- T4 cannot start at 2:00.

We can model this scheduling problem by creating five variables, one for each task, where each variable has the domain {1:00, 2:00, 3:00}. The corresponding constraint graph is shown in Figure 2.3, and the relations expressed by the graph are shown beside the figure.[2]

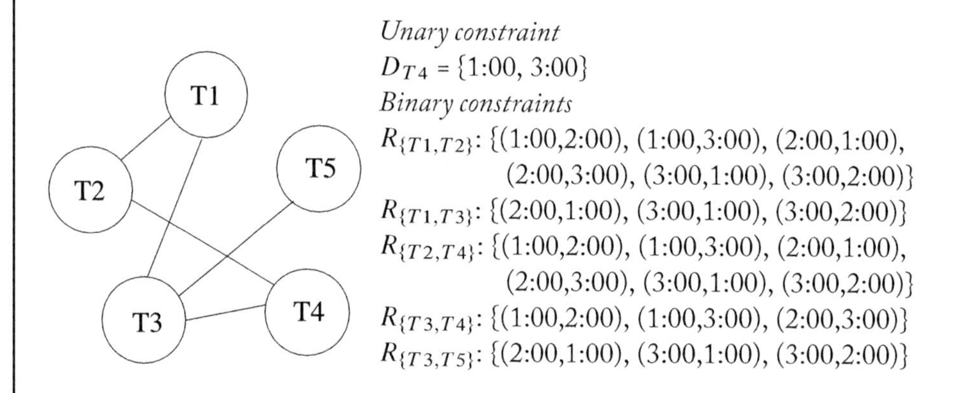

Figure 2.3: The constraint graph and constraint relations of the scheduling problem.

Sometimes we express the relation R_i as a cost function C_i, where $C(\mathbf{x}_{S_i}) = 1$ if $\mathbf{x}_{S_i} \in R_i$ and 0 otherwise. In this case the combination operator is a product. We will switch between these two alternative specification as needed. If we want to count the number of solutions we merely change the marginalization operator to be summation. If on the other hand we want merely to query whether the constraint network has a solution, we can let the marginalization operator be maximization. We let $\mathbf{Z} = \{\emptyset\}$, so that the summation occurs over all the variables. We will get "1" if the constraint problem has a solution and "0" otherwise.

Propositional Satisfiability One special case of the constraint satisfaction problem is what is called *propositional satisifiability* (usually referred to as SAT). Given a formula φ in *conjunctive normal form* (CNF), the SAT problem is to determine whether there is a truth-assignment of values to its variables such that the formula evaluates to true. A formula is in conjunctive normal form

[2]Example taken from [Dechter, 2003].

if it is a conjunction of *clauses* $\alpha_1, \ldots, \alpha_t$, where each clause is a disjunction of *literals* (propositions or their negations). For example, $\alpha = (P \vee \neg Q \vee \neg R)$ and $\beta = (R)$ are both clauses, where P, Q, and R are propositions, and P, $\neg Q$, and $\neg R$ are literals. $\varphi = \alpha \wedge \beta = (P \vee \neg Q \vee \neg R) \wedge (R)$ is a formula in conjunctive normal form.

Propositional satisfiability can be defined as a constraint satisfaction problem in which each proposition is represented by a variable with domain {0, 1}, and a clause is represented by a constraint. For example, the clause $(\neg A \vee B)$ is a relation over its propositional variables that allows all tuple assignments over (A, B) except $(A = 1, B = 0)$.

2.4 COST NETWORKS

In constraint networks, the local functions are constraints, i.e., functions that assign a Boolean value to any assignment in its domain. However, it is straightforward to extend constraint networks to accommodate real-valued relations using a graphical model called a *cost network*. In cost networks, the local functions represents cost-components, and the sum of these cost-components is the global cost function of the network. The primary task is to find an assignment of the variables such that the global cost function is optimized (minimized or maximized). Cost networks enable one to express preferences among local assignments and, through their global costs to express preferences among full solutions.

Often, problems are modeled using both constraints and cost functions. The constraints can be expressed explicitly as being functions of a different type than the cost functions, or they can be modeled as cost components themselves. It is straightforward to see that cost networks are graphical model where the combination operator is summation.

Definition 2.15 Cost network, combinatorial optimization. A *cost network* is a 4-tuple graphical model, $C = \langle \mathbf{X}, \mathbf{D}, \mathbf{F}, \sum \rangle$, where $\mathbf{X}$ is a set of variables $\mathbf{X} = \{X_1, \ldots, X_n\}$, associated with a set of discrete-valued domains, $\mathbf{D} = \{D_1, \ldots, D_n\}$, and a set of local cost functions $\mathbf{F} = \{f_{S_1}, \ldots, f_{S_r}\}$. Each f_{S_i} is a real-valued function (called also cost-component) defined on a subset of variables $\mathbf{S}_i \subseteq \mathbf{X}$. The local cost components are combined into a global cost function via the $\sum$ operator. Thus, the cost network represents the function

$$C(\mathbf{x}) = \sum_i f_{\mathbf{S}_i}(\mathbf{x}_{\mathbf{S}_i}) \; or$$

in simplified notations as

$$C(\mathbf{x}) = \sum_{f \in F} f(\mathbf{x}_f) \, .$$

The primary optimization task (which we will assume to be a minimization, w.l.o.g) is to find an optimal solution for the global cost function $F = \sum_i f_i$. Namely, finding a tuple $\mathbf{x}$ such that $\mathbf{x} = argmin_{\mathbf{X}} \sum_i f_{\mathbf{S}_i}(\mathbf{x}_{\mathbf{S}_i})$. We can associate the cost model with its primal graph in the usual way.

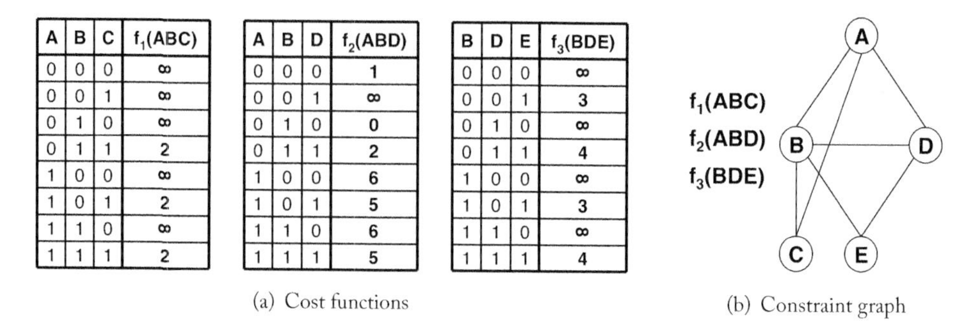

A	B	C	$f_1(ABC)$
0	0	0	∞
0	0	1	∞
0	1	0	∞
0	1	1	2
1	0	0	∞
1	0	1	2
1	1	0	∞
1	1	1	2

A	B	D	$f_2(ABD)$
0	0	0	1
0	0	1	∞
0	1	0	0
0	1	1	2
1	0	0	6
1	0	1	5
1	1	0	6
1	1	1	5

B	D	E	$f_3(BDE)$
0	0	0	∞
0	0	1	3
0	1	0	∞
0	1	1	4
1	0	0	∞
1	0	1	3
1	1	0	∞
1	1	1	4

(a) Cost functions (b) Constraint graph

Figure 2.4: A cost network.

Weighted Constraint Satisfaction Problems A special class of cost networks that has gained considerable interest in recent years is a graphical model called the Weighted Constraint Satisfaction Problem (WCSP) [Bistarelli *et al.*, 1997]. These networks extends the classical constraint satisfaction problem formalism with *soft constraints*, that is, positive integer-valued local cost functions.

Definition 2.16 WCSP. A *Weighted Constraint Satisfaction Problem (WCSP)* is a graphical model $\langle \mathbf{X}, \mathbf{D}, \mathbf{F}, \sum \rangle$ where each of the functions $f_i \in \mathbf{F}$ assigns "0" (no penalty) to allowed tuples and a positive integer penalty cost to the forbidden tuples. Namely, $f_i : D_{\mathbf{S}_i} \rightarrow \mathbb{N}$, where S_i is the scope of the function.

Many real-world problems can be formulated as cost networks and often fall into the weighted CSP class. This includes resource allocation problems, scheduling [Bensana *et al.*, 1999], bioinformatics [de Givry *et al.*, 2005; Thébault *et al.*, 2005], combinatorial auctions [Dechter, 2003; Sandholm, 1999], and maximum satisfiability problems [de Givry *et al.*, 2003].

Example 2.17 Figure 2.4 shows an example of a WCSP instance with Boolean variables. The cost functions are given in Figure 2.4(a), and the associated graph is shown in Figure 2.4(b). Note that a value of ∞ in the cost function denotes a hard constraint (i.e., high penalty). You should verify that the minimal cost solution of the problem is 5, which corresponds to the assignment $(A = 0, B = 1, C = 1, D = 0, E = 1)$.

The task of MAX-CSP, namely of finding a solution that satisfies the maximum number of constraints (when the problem is inconsistent), can be formulated as a cost network by treating each relation as a cost function that assigns "0" to consistent tuples and "1" otherwise. Since all violated constraints are penalized equally, the global cost function will simply count the number of violations. In this case the combination operator is summation and the marginalization operator is minimization. Namely, the task is to find $\Downarrow_{\emptyset} \bigotimes_i f_{\mathbf{S}_i}$, namely, to find,$argmin_{\mathbf{X}}(\sum_i f_{\mathbf{S}_i})$.

Definition 2.18 MAX-CSP. A *MAX-CSP* is a WCSP $\langle \mathbf{X}, \mathbf{D}, \mathbf{F} \rangle$ with all penalty costs equal to 1. Namely, $\forall f_i \in \mathbf{F}$, $f_i : D_{f_i} \to \{0, 1\}$.

Maximum Satisfiability. In the same way that propositional satisfiability (SAT) can be seen as a constraint satisfaction problem over logical formulas in conjunctive normal form, so can the problem of *maximum satisfiability* (MAX-SAT) be formulated as a MAX-CSP problem. In this case, given a set of Boolean variables and a collection of clauses defined over subsets of those variables, the goal is to find a truth assignment that violates the least number of clauses. Naturally, if each clause is associated with a positive weight, then the problem can be described as a WCSP. The goal of this problem, called *weighted maximum satisfiability* (weighted MAX-SAT), is to find a truth assignment such that the sum weight of the violated clauses is minimized.

Integer Linear Programs. Another well-known class of optimization task is integer linear programming. It is formulated over variables that can be assigned integer values (finite or infinite). The task is to find an optimal solution to a linear cost function $F(x) = \sum_i \alpha_i x_i$ that satisfies a set of linear constraints.

Definition 2.19 Integer linear programming. An *Integer Linear Programming Problem (ILP)* is a graphical model $\langle \mathbf{X}, \mathbf{N}, \mathbf{F} = \{f_1, ... f_n, C_1, ..., C_l\}, \sum \rangle$ having two types of functions. Linear cost components $f_i(x_i) = \alpha_i x_i$ for each variable X_i, where α_i is a real number. The scopes are singleton variables. The constraints are of weighted csp type, each defined on scope $\mathbf{S}_i$. They are specified by

$$C_i(\mathbf{x}_{S_i}) = \left\{ 0, \quad if \ \sum_{x_j \in S_i} \lambda_{i_j} \cdot x_j \leq \lambda_i \right.$$

or infinity otherwise. The λ's are given real-valued constants. The marginalization operator is minimization or maximization.

2.5 PROBABILITY NETWORKS

As mentioned previously, Bayesian networks and Markov networks are the two primary formalisms for expressing probabilistic information via graphical models.

2.5.1 BAYESIAN NETWORKS

A *Bayesian network* [Pearl, 1988] is defined by a directed acyclic graph over vertices that represent random variables of interest (e.g., the temperature of a device, gender of a patient, feature of an object, occurrence of an event). The arc from one node to another is meant to signify a direct causal influence or correlation between the respective variables, and this influence is quantified by the conditional probability of the child variable given all of its parents variables. Therefore, to define a Bayesian network, one needs both a directed graph and the associated conditional probability functions. To be consistent with our graphical models description we define Bayesian network as follows.

Definition 2.20 (Bayesian networks) A *Bayesian network (BN)* is a 4-tuple $\mathcal{B} = \langle \mathbf{X}, \mathbf{D}, \mathbf{P}_G, \prod \rangle$. $\mathbf{X} = \{X_1, \ldots, X_n\}$ is a set of ordered variables defined over domains $\mathbf{D} = \{D_1, \ldots, D_n\}$, where $o = (X_1, \ldots, X_n)$ is an ordering of the variables. The set of functions $\mathbf{P}_G = \{P_1, \ldots, P_n\}$ consist of Conditional Probability Tables (CPTs for short) $P_i = \{P(X_i | \mathbf{Y}_i)\}$ where $\mathbf{Y}_i \subseteq \{X_{i+1}, ..., X_n\}$. These P_i functions can be associated with a directed acyclic graph G in which each node represents a variable X_i and there is a directed arc from each parent variable of X_i to X_i. The Bayesian network $\mathcal{B}$ represents the probability distribution over $\mathbf{X}$, $P_{\mathcal{B}}(\mathbf{x}) = \prod_{i=1}^{n} P(x_i | \mathbf{x}_{pa(X_i)})$ where $pa(X)$ are the parents of X in G. We define an evidence set e as an instantiated subset of evidence variables $\mathbf{E}$. The Bayesian network always yields a valid joint probability distribution.

Moreover, it is consistent with its input CPTs. Namely, for each X_i and its parent set $\mathbf{Y}_i$, it can be shown that

$$P_{\mathcal{B}}(X_i | \mathbf{Y}_i) \propto \sum_{\mathbf{X} - \{X_i \cup \mathbf{Y}_i\}} P_{\mathcal{B}}(\mathbf{x}) = P(X_i | \mathbf{Y}_i).$$

where the last summand on the right is the input CPT for variable X_i.

Therefore a Bayesian network is a graphical model, where the combination operator is product, $\bigotimes = \prod$. The primal graph of a Bayesian network is called a moral graph and it connects any two variables appearing in the same CPT. The moral graph can also be obtained from the directed graph G by connecting all the parents of each child node and making all directed arcs undirected.

Example 2.21 [Pearl, 1988] Figure 2.5(a) is a Bayesian network over six variables, and Figure 2.5(b) shows the corresponding moral graph. The example expresses the causal relationship between variables "season" (A), "the automatic sprinkler system is on" (B), "whether it rains or does not rain" (C), "manual watering is necessary" (D), "the wetness of the pavement" (F), and "the pavement is slippery" (G). The Bayesian network is defined by six conditional probability tables each associated with a node and its parents. For example, the CPT of F describes the probability that the pavement is wet ($F = 1$) for each status combination of the sprinkler and raining. Possible CPTs are given in Figure 2.5(c).

The conditional probability tables contain only half of the entries because the rest of the information can be derived based on the property that all the conditional probabilities sum to

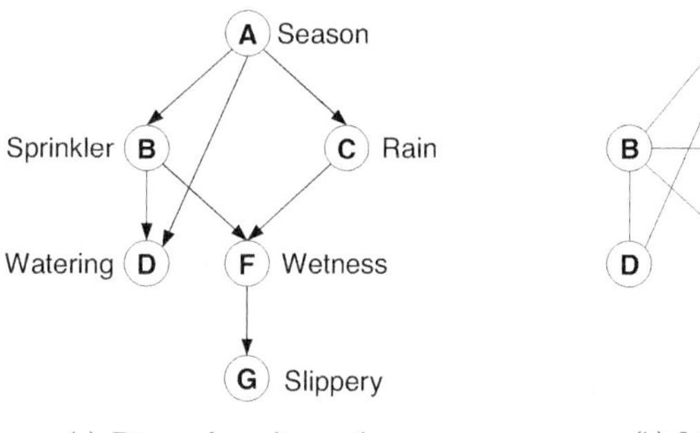

(a) Directed acyclic graph

(b) Moral graph

B	C	F	$P(F\|B,C)$
false	false	true	0.1
true	false	true	0.9
false	true	true	0.8
true	true	true	0.95

B	A = *winter*	D	$P(D\|A,B)$
false	false	true	0.3
true	false	true	0.9
false	true	true	0.1
true	true	true	1

A	C	$P(C\|A)$
Summer	true	0.1
Fall	true	0.4
Winter	true	0.9
Spring	true	0.3

A	B	$P(B\|A)$
Summer	true	0.8
Fall	true	0.4
Winter	true	0.1
Spring	true	0.6

F	G	$P(G\|F)$
false	true	0.1
true	true	1

(c) Possible CPTs that accompany our example

Figure 2.5: Belief network $P(G,F,C,B,A) = P(G|F)P(F|C,B)P(D|A,B)P(C|A)P(B|A)P(A)$.

1. This Bayesian network expresses the probability distribution $P(A,B,C,D,F,G) = P(A) \cdot P(B|A) \cdot P(C|A) \cdot P(D|B,A) \cdot P(F|C,B) \cdot P(G|F)$.

Next, we define the main queries over Bayesian networks.

Definition 2.22 (Queries over Bayesian networks) Let $\mathcal{B} = \langle \mathbf{X}, \mathbf{D}, \mathbf{P}_G, \prod \rangle$ be a Bayesian network. Given evidence $E = e$ where E is the evidence variables and e is their assignment, the primary queries over Bayesian networks are to find the following quantities.

1. **Posterior marginals, or belief updating.** For every X_i not in E the belief is defined by $bel(X_i) = P_{\mathcal{B}}(X_i|e)$.

$$P(X_i|e) = \sum_{\mathbf{X}-X_i} \prod_j P(X_j|X_{pa_j}, e)$$

2. The **probability of evidence** is $P_{\mathcal{B}}(E = e)$. Formally,

$$P_{\mathcal{B}}(E = e) = \sum_{\mathbf{X}} \prod_j P(X_j|X_{pa_j}, e)$$

3. The **most probable explanation (*mpe*)** is an assignment $\mathbf{x}^o = (x^o{}_1, ..., x^o{}_n)$ satisfying

$$\mathbf{x}^o = argmax_{\mathbf{X}} \mathbf{P}_{\mathcal{B}} = argmax_{\mathbf{X}} \prod_j P(X_j|X_{pa_j}, e).$$

The *mpe* value is $\mathbf{P}_{\mathcal{B}}(\mathbf{x}^o)$, sometime also called MAP.

4. **Maximum a posteriori hypothesis (marginal *map*).** Given a set of hypothesized variables $\mathbf{A} = \{A_1, ..., A_k\}$, $\mathbf{A} \subseteq \mathbf{X}$, the *map* task is to find an assignment $\mathbf{a}^o = (a^o{}_1, ..., a^o{}_k)$ such that

$$\mathbf{a}^o = argmax_{\mathbf{A}} \sum_{\mathbf{X}-\mathbf{A}} \mathbf{P}(\mathbf{X}|\mathbf{e}) = argmax_{\mathbf{A}} \sum_{\mathbf{X}-\mathbf{A}} \prod_j P(X_j|X_{pa_j}, e)$$

These queries are applicable to a variety of applications such as situation assessment, diagnosis, probabilistic decoding and linkage analysis, to name a few. To answer the above queries over $\mathcal{B} = \langle \mathbf{X}, \mathbf{D}, \mathbf{P}_G, \prod \rangle$ we use as marginalization operators either summation or maximization. In particular, the query of finding the probability of the evidence can be expressed as $\Downarrow_{\emptyset} \bigotimes_i f_i = \sum_X \prod_i P_i$. The *belief updating* task, when given evidence e, can be formulated using the summation as a marginalization operator, where $\mathbf{Z}_i = \{X_i\}$. Namely, $\forall X_i, P_{\mathcal{B}}(X_i|e) = \Downarrow_{X_i} \bigotimes_k f_k = \sum_{\{X-X_i;E=e\}} \prod_k P_k$. The mpe task is defined by a maximization operator where $\mathbf{Z} = \{\emptyset\}$, yielding *mpe* defined by $\Downarrow_{\emptyset} \bigotimes_i f_i = \max_X \prod_i P_i$. If we want to get the actual mpe assignment we would need to use the *argmax* operator. Finally, the marginal map is defined by both summation and maximization.

2.5.2 MARKOV NETWORKS

Markov networks, also called *Markov Random Fields (MRF)*, are undirected probabilistic graphical models very similar to Bayesian networks. However, unlike Bayesian networks they convey undirectional information, and are therefore defined over an undirected graph. Moreover, whereas the functions in Bayesian networks are restricted to be conditional probability tables of children given their parents in some directed graph, in Markov networks the local functions, called potentials, can be defined over any subset of variables. These potential functions between random variables can be thought of as expressing some kind of a correlation information. When a configuration to a subset of variables is likely to occur together their potential value may be large. For instance in vision scenes, variables may represent the grey levels of pixels, and neighboring pixels are likely to have similar grey values. Therefore, they can be given a higher potential level. Other applications of Markov networks are in physics (e.g., modeling magnetic behaviors of crystals). They convey symmetrical information and can be viewed as the probabilistic counterpart of constraint or cost networks, whose functions are symmetrical as well.

Like a Bayesian network, a Markov network also represents a joint probability distribution, even though its defining local functions do not have a clear probabilistic semantics. In particular, they do not express local marginal probabilities (see [Pearl, 1988] for a discussion).

Definition 2.23 Markov networks. A Markov network is a graphical model $\mathcal{M} = \langle \mathbf{X}, \mathbf{D}, \mathbf{H}, \prod \rangle$ where $\mathbf{H} = \{\psi_1, \ldots, \psi_m\}$ is a set of potential functions where each potential ψ_i is a non-negative real-valued function defined over a scope of variables $\mathcal{S} = \{\mathbf{S}_1, ..., \mathbf{S}_m\}$. $\mathbf{S}_i$. The Markov network represents a global joint distribution over the variables $\mathbf{X}$ given by:

$$P_{\mathcal{M}} = \frac{1}{Z} \prod_{i=1}^{m} \psi_i \quad , \quad Z = \sum_{\mathbf{X}} \prod_{i=1}^{m} \psi_i$$

where the normalizing constant Z is called the partition function.

Queries. The primary queries over Markov networks are the same as those of Bayesian network. That is, computing the posterior marginal distribution over all variables $X_i \in \mathbf{X}$, finding the *mpe* value and a corresponding assignment (configuration) and finding the partition function. It is not hard to see that this later query is mathematically identical to computing the probability of evidence. Like Bayesian networks, Markov networks are graphical models whose combination operator is the product operator, $\bigotimes = \prod$ and the marginalization operator can be summation, or maximization, depending on the query.

Example 2.24

Figure 2.6 shows a 3×3 square grid Markov network with 9 variables $\{A, B, C, D, E, F, G, H, I\}$. The 12 potentials are: $\psi_1(A, B)$, $\psi_2(B, C)$, $\psi_3(A, D)$, $\psi_4(B, E)$, $\psi_5(C, F)$, $\psi_6(C, D)$, $\psi_7(D, E)$, $\psi_8(D, G)$, $\psi_9(E, H)$, $\psi_{10}(F, I)$, $\psi_{11}(G, H)$, and $\psi_{12}(H, I)$. The Markov network represents the probability distribution formed by taking a product of these twelve functions and then normalizing. Namely, given that $\mathbf{x} = (a, b, c, d, e, f, g, h, i)$

$$F(a, b, c, d, e, f, g, h, i) \propto$$

$$\psi_1(a, b) \cdot \psi_2(b, c) \cdot \psi_3(a, d) \cdot \psi_4(b, e) \cdot \psi_5(c, f) \cdot \psi_6(d, e) \cdot \psi_7(e, f) \cdot \psi_8(d, g)$$
$$\cdot \psi_9(e, h) \cdot \psi_{10}(f, i) \cdot \psi_{11}(g, h) \cdot \psi_{12}(h, I)$$

where $Z = \sum_{a,b,c,d,e,f,g,h,i} F(a, b, c, d, e, f, g, h, i)$ is the partition function.

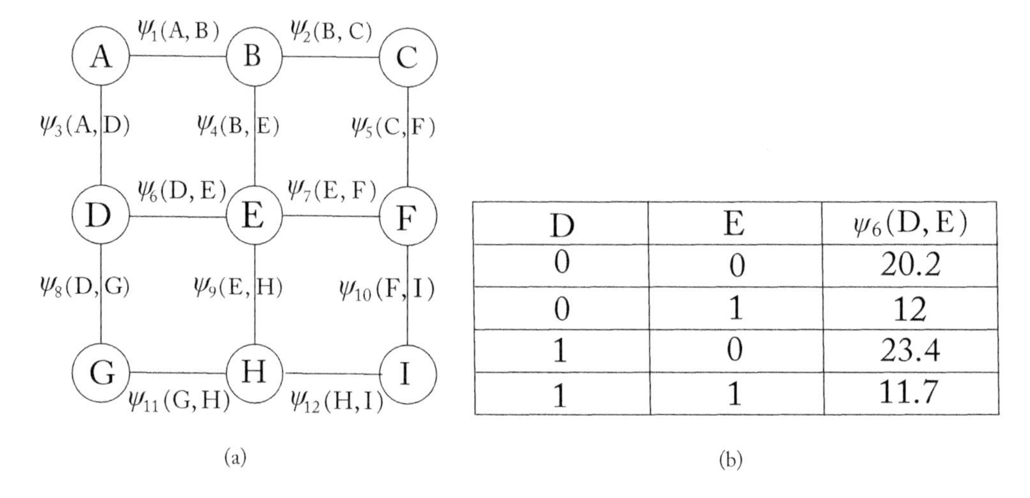

D	E	$\psi_6(D, E)$
0	0	20.2
0	1	12
1	0	23.4
1	1	11.7

Figure 2.6: (a) An example 3×3 square grid Markov network (ising model) and (b) an example potential $H_6(D, E)$.

Markov networks typically are generated by starting with a graph model which describes the variables of interest and how they depend on each other, like in the case of image analysis whose graph is a grid. Then the user defines potential functions on the cliques of the graph. A well-known example is the *ising model*. This model arise from statistical physics [Murphy, 2012]. It was used to model the behavior of magnets. The structure is a grid, where the variables have values $\{-1, +1\}$. The potential express the desire to have neighboring variables have the same value. The resulting Markov network is called a Markov Random Field (MRF). Alternatively, like in the case of constraint networks, if the potential functions are specified with no explicit reference to a graph (perhaps representing some local probabilistic information or compatibility information) the graph emerges as the associated primal graph.

Markov networks provide some more freedom from the modeling perspective, allowing to express potential functions on any subset of variables. This, however, comes at the cost of loosing semantic clarity. The meaning of the input local functions relative to the emerging probability distribution is not coherent. In both Bayesian networks and Markov networks the modeling process starts from the graph. In the Bayesian network case the graph restricts the CPTs to be defined for each node and its parents. In Markov networks, the potentials should be defined on the maximal cliques. For more see [Pearl, 1988].

2.6 MIXED NETWORKS

In this section, we introduce the mixed network, a graphical model which allows both probabilistic information and deterministic constraints and which provides a coherent meaning to the combination.

Definition 2.25 Mixed networks. Given a belief network $\mathcal{B} = \langle \mathbf{X}, \mathbf{D}, \mathbf{P}_G, \prod \rangle$ that expresses the joint probability $P_\mathcal{B}$ and given a constraint network $\mathcal{R} = \langle \mathbf{X}, \mathbf{D}, \mathbf{C}, \bowtie \rangle$ that expresses a set of solutions denoted ρ, a mixed network based on $\mathcal{B}$ and $\mathcal{R}$ denoted $\mathcal{M}_{(\mathcal{B},\mathcal{R})} = \langle \mathbf{X}, \mathbf{D}, \mathbf{P}, \mathbf{C} \rangle$ is created from the respective components of the constraint network and a Bayesian network as follows: the variables $\mathbf{X}$ and their domains are shared (we could allow non-common variables and take the union), and the functions include the CPTs in $\mathbf{P}_G$ and the constraints in $\mathbf{C}$. The mixed network expresses the conditional probability $P_\mathcal{M}(\mathbf{X})$:

$$P_\mathcal{M}(\mathbf{x}) = \begin{cases} P_\mathcal{B}(\mathbf{x} \mid \mathbf{x} \in \rho), & if \ \ \mathbf{x} \in \rho \\ 0, & otherwise. \end{cases}$$

Example 2.26 Consider a scenario involving social relationship between three individuals Alex (A), Becky (B), and Chris (C). We know that if Alex goes to a party Becky will go, and if Chris goes Alex goes. We also know the weather effects these three individuals differently and they will or will not go to a party with some differing likelihood. We can express the relationship between going to the party and the weather using a Bayesian network (Figure 2.7a), while the social relationship using a propositional formula (see Figure 2.7b).

The mixed network have two types of functions: probabilistic local functions and constraints. This is a graphical model whose combination operator is product, when we assume that constraints have their cost-based representation.

Queries. Posterior marginals, *mpe* and *marginal map* queries over probabilistic networks can be extended to mixed networks straight-forwardly. They are well-defined relative to the mixed probability distribution $P_\mathcal{M}$. Since $P_\mathcal{M}$ is not well defined for *inconsistent* constraint networks we always assume that the constraint network portion is consistent.

Mixed networks give rise to a new query, which is to find the probability of a consistent tuple, namely, we want to determine $P_\mathcal{B}(\mathbf{x} \in sol(\mathcal{R}))$. We will call this *a Constraint Probability Evaluation (CPE)*. Note that evidence is a special type of constraint. So this is an extension of probability of the evidence, or the partition function query.

Definition 2.27 Queries on mixed networks. We consider the following two new queries.

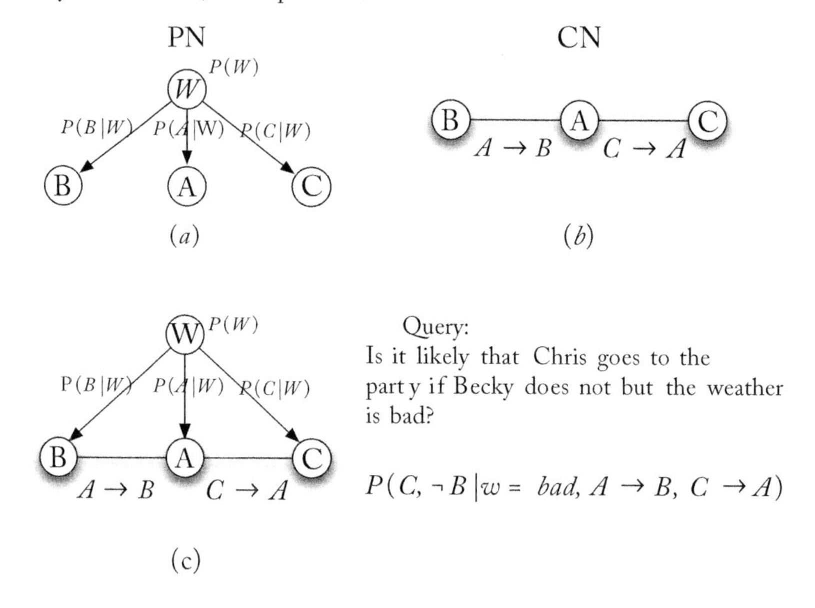

Figure 2.7: The part example, a Bayesian network (a), constraint formula (b), and a mixed network (c).

- Given a mixed network $\mathcal{M}_{(\mathcal{B},\mathcal{R})}$, where $\mathcal{B} = \langle \mathbf{X}, \mathbf{D}, \mathbf{P}, \prod \rangle$ and $\mathcal{R} = \langle \mathbf{X}, \mathbf{D}, \mathbf{C} \rangle$ the *constraint* Probability Evaluation (CPE) task is to find the probability $P_{\mathcal{B}}(\mathbf{x} \in sol(\mathcal{R}))$. If R is a CNF expression φ, the cnf probability evaluation seeks $P_{\mathcal{B}}(\mathbf{x} \in Mod(\varphi))$, where $Mod(\varphi)$ are the models (solutions of φ).

- *Belief assessment of a constraint or of a CNF expression* is the task of assessing $P_{\mathcal{B}}(X|\varphi)$ for every variable X. Since $P(X_i|\varphi) = \alpha \cdot P(X_i \wedge \varphi)$ where α is a normalizing constant relative to X, computing $P_{\mathcal{B}}(X|\varphi)$ reduces to a CPE task over $\mathcal{B}$ for the query $((X_i = x_i) \wedge \varphi)$. In other words we want to find $P_B(\mathbf{x}|X_i = x_i, \mathbf{x} \in Mod(\varphi))$. More generally, $P_{\mathcal{B}}(\varphi|\psi) = \alpha_{\varphi} \cdot P_{\mathcal{B}}(\varphi \wedge \psi)$ where α_{φ} is a normalization constant relative to all the models of φ.

The problem of evaluating the probability of CNF queries over Bayesian networks has various applications. One example is network reliability: Given a communication graph with a source and a destination, one seeks to diagnose the failure of communication. Since several paths may be available between source and destination, the failure condition can be described by a CNF formula as follows. Failure means that for all paths (conjunctions) there is a link on that path (disjunction) that fails. Given a probabilistic fault model of the network, the task is to assess the probability of a failure [Portinale and Bobbio, 1999]. There are many examples in modeling travel patterns of human and in natural language processing. [Chang *et al.*, 2012; Gogate *et al.*, 2005].

We conclude with some example queries over mixed networks.

Definition 2.28 The weighted counting task. Given a mixed network $\mathcal{M} = \langle \mathbf{X}, \mathbf{D}, \mathbf{P}_G, \mathbf{C} \rangle$, where $\mathbf{P}_G = \{P_1, ..., P_m\}$ the weighted counting task is to compute the normalization constant given by:

$$Z = \sum_{\mathbf{x} \in Sol(\mathbf{C})} \prod_{i=1}^{m} P_i \ . \tag{2.1}$$

Equivalently, if we have a cost-based representation of the constraints in C as 0/1 functions, we can rewrite Z as:

$$Z = \sum_{\mathbf{X}} \prod_{i=1}^{m} P_i \prod_{j=1}^{p} C_j \ . \tag{2.2}$$

We will refer to Z as **weighted counts** and we can see that mathematically, it is identical to the partition function.

Definition 2.29 Marginal task, belief updating. Given a mixed network $\mathcal{M} = \langle X, D, P, C \rangle$, where $P = \{P_1, ..., P_n\}$ and $\mathcal{R} = \langle \mathbf{X}, \mathbf{D}, \mathbf{C} \rangle$, the marginal task is to compute the marginal distribution at each variable. Namely, for each variable X_i compute:

$$\mathbf{P}_{\mathcal{M}}(x_i) = \sum_{\mathbf{X}} \delta_{X_i} \mathbf{P}_{\mathcal{M}}, \text{ where } \delta_{X_i}(\mathbf{x}) = \begin{cases} 1 & \text{if } X_i \text{ is assigned the value } x_i \\ 0 & \text{otherwise.} \end{cases}$$

When we are given a probabilistic network that has zeros, we can extract a constraint portion from it, generating an explicit mixed network as we show below.

It is easy to see that the weighted counts over a mixed network specialize to (a) the probability of evidence in a Bayesian network, (b) the partition function in a Markov network, and (c) the number of solutions of a constraint network. The marginal problem can express the posterior marginals in a Bayesian or Markov network.

2.7 SUMMARY AND BIBLIOGRAPHICAL NOTES

The work on graphical models can be seen as originating from two communities. The one that centers on statistics and probabilities and aims at capturing probability distributions vs. the one that centers on deterministic relationships, such as constraint networks and logic systems. Each represents an extreme point in a spectrum of models. Each went through the process of generalization and extensions towards the other; probabilistic models were augmented with constraint processing and utility information (e.g., leading to influence diagrams), and constraint networks were extended to soft constraints and into fuzzy type information.

The seminal work by Bistareli et al. [Bistarelli *et al.*, 1997] provides a foundational unifying treatment of graphical models, using the mathematical framework of semirings. Various semirings yield different graphical models, using the umbrella name Soft Constraints. The work emerged from and generalizes the area of constraint networks. Constraint networks were distinguished as semirings that are *idempotent*. For a complete treatment, see [Bistarelli, 2004]. Another line of work rooted at probabilistic networks was introduced by Shenoy and Shafer who provide an axiomatic treatment for probability and belief-function propagation [Shafer and Shenoy, 1990; Shenoy, 1992]. Their framework is focused on an axiomatic formulation of the two operators of combination and marginalization in graphical models. The work by Dechter [Dechter, 1996a, 1999], focusing on unifying variable elimination algorithms, demonstrates that common algorithms can be applied across various graphical models such as constraints networks, cost-networks, propositional cnfs, influence diagrams, and probabilistic networks and that it can be expressed using also the two operation of combination and marginalization [Kask and Dechter, 2005]. This work is the basis of the exposition in this book. Other related work focusing on message-passing perspective over certain restricted graphs is Srinivas and McEliece [Aji and McEliece, 2000].

CHAPTER 3

Inference: Bucket Elimination for Deterministic Networks

This chapter is the first of two chapters in which we introduce the *bucket-elimination* inference scheme. It is a variable elimination scheme that generalizes dynamic programming and characterizes all inference algorithms over graphical models. As noted, by *inference* we mean algorithms that solve queries by inducing equivalent model representations according to some set of inference rules. Bucket-elimination algorithms are *knowledge-compilation*, also called *reparameterization schemes* methods: they generate an equivalent representation of the input problem from which various queries are answerable in polynomial time. We will see that the bucket-elimination scheme is applicable to most, if not all, of the types of queries and graphical models we discussed in Chapter 2, but its general structure and properties are most readily understood in the context of constraint networks. In this chapter, the primary query is whether or not an input constraint network is consistent. In the following chapter, we will apply this scheme to probabilistic reasoning and combinatorial optimization.

To illustrate the basic idea behind bucket elimination, let's walk through a simple constraints problem. Consider the graph coloring problem in Figure 3.1a. The task is to assign one of two colors (green or red) to each node in the graph so that adjacent nodes will have different colors. Here is one way to solve this problem: consider node E first. It can be colored either green or red. Since only two colors are available it follows that D and C must have identical colors; thus, $C = D$ can be inferred, and we can add this as a new constraint to our network without changing its solutions set. We can ignore variable E from now on since we already summarized its impact on the rest of the problem when we added $C = D$. We focus on variable C next. Together, the inferred constraint $C = D$ and the input constraint $C \neq B$ imply that $D \neq B$, and we add this constraint to the model. Having taken into account the effect of C on the other variables in the network, we can ignore C also from now on. Continuing in this fashion with node D, we infer $A = B$. However, since there is an input constraint $A \neq B$ we have reached a contradiction and can conclude that the original set of constraints is inconsistent.

The algorithm which we just executed, is known as *adaptive-consistency* in the constraint literature [Dechter and Pearl, 1987] and it can solve any constraint satisfaction problem. The algorithm works by processing and *eliminating* variables one by one, while deducing the effect of the eliminated variable on the rest of the problem. The elimination operation first *joins* all the

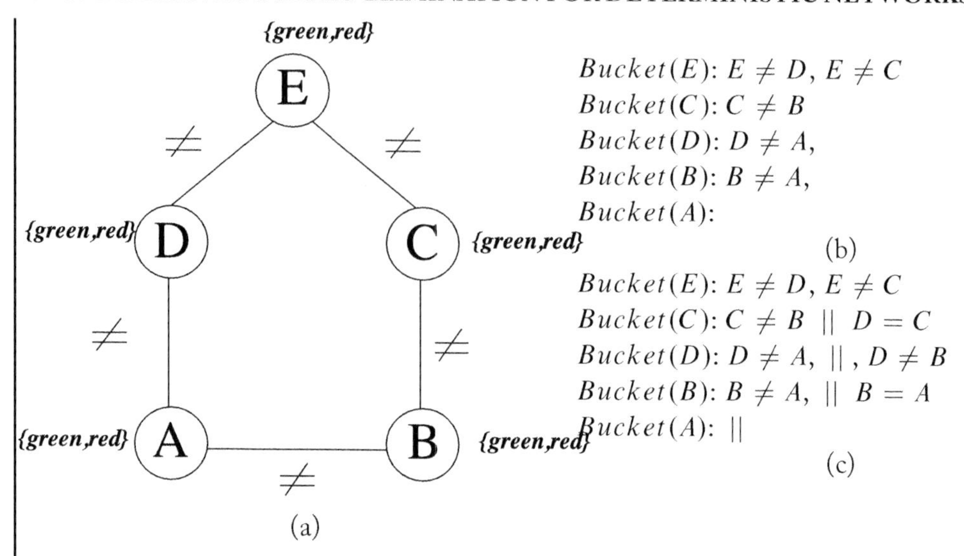

Figure 3.1: A graph coloring example (a) and a schematic execution of adaptive-consistency (b,c).

relations that are defined on the current variable and then projects out the variable. Adaptive-consistency can be described using a data structure called *buckets* as follows: given an ordering of the variables, we process the variables from last to first. In the previous example, the ordering was $d = A, B, D, C, E$, and we processed the variables from E to A. Note that we will use this convention throughout: we assume that the inference algorithm process the variables from last to first w.r.t to a given ordering. (The reason for that will be clear later.) The first step is to partition the constraints into *ordered buckets*, so that the bucket for the current variable contains all constraints that mention the current variable and that have not been already placed in a previous bucket. In our example, all the constraints mentioning the last variable E are put in a bucket designated as $bucket_E$. Subsequently, all the remaining constraints mentioning D are placed in $bucket_D$, and so on. The initial partitioning of the constraints is depicted in Figure 3.1b. The general partition rule is that each constraint identifies the variable in its scope that appears latest in the ordering, and then places the constraint in the bucket of the identified variable.

After this initialization step, the buckets are processed from last to first. Processing a bucket means solving the subproblem defined by the constraints in the bucket and then inferring the constraint that is imposed by that subproblem on the rest of the variables. In other words, we compute the constraint that the bucket-variable induces on the variables that precede it in the ordering. As we saw, processing bucket E produces the constraint $D = C$, which is placed in $bucket_C$. By processing $bucket_C$, the constraint $D \neq B$ is generated and placed in $bucket_D$. While processing bucket D, we generate the constraint $A = B$ and put it in $bucket_B$. When processing $bucket_B$, inconsistency is discovered between the inferred $A \neq B$ and the input constraint $A = B$. The buckets' final contents are shown in Figure 3.1c. The new inferred constraints are displayed to the right of the bar in each bucket.

Observe that because the new added constraints are inferred, the problem itself does not change in the sense that with or without the added constraints it has the same set of solutions. However, what is significant is that once all the buckets are processed, and if no inconsistencies were discovered, a solution can be generated in a *backtrack-free* manner. This means that a solution can be assembled by assigning values to the variables progressively, starting with the first variable in ordering d while respecting all the current constraints in a bucket. This process is guaranteed to continue until all the variables are assigned a value from their respective domains, thus yielding a *solution* to the problem. The notion of *backtrack-free* constraint network relative to an ordering is central to the theory of constraint processing and will be defined shortly.

3.1 BUCKET-ELIMINATION FOR CONSTRAINT NETWORKS

We have presented an informal definition of the bucket-elimination algorithm on constraint networks called adaptive-consistency. Here we will provide a formal definition of the algorithm, using the formalism of constraint networks introduced in the previous chapter and utilizing the following operations:

Definition 3.1 Operations on constraints: select, project, join. Let R be a relation on a set $\mathbf{S}$ of variables, let $\mathbf{Y} \subseteq \mathbf{S}$ be a subset of the variables, and let $\mathbf{y}$ be an instantiation of the variables in $\mathbf{Y}$. We denote by $\sigma_{\mathbf{y}}(R)$ the selection of those tuples in $\mathcal{R}$ that agree with $\mathbf{Y} = \mathbf{y}$. We denote by $\pi_{\mathbf{Y}}(R)$ the projection of the relation R on the subset Y, that is, a tuple $\mathbf{Y} = y$ appears in $\pi_{\mathbf{Y}}(R)$ if and only if it can be extended to a full tuple in R. Let $R_{\mathbf{S}_1}$ be a relation on a set $\mathbf{S}_1$ of variables and let $R_{\mathbf{S}_2}$ be a relation on a set $\mathbf{S}_2$ of variables. We denote by $R_{\mathbf{S}_1} \bowtie R_{\mathbf{S}_2}$ the join of the two relations. The join of $R_{\mathbf{S}_1}$ and $R_{\mathbf{S}_2}$ is a relation defined over $\mathbf{S}_1 \cup \mathbf{S}_2$ containing all and only the tuples $\mathbf{t}$, satisfying $\mathbf{t}_{\mathbf{S}_1} \in R_{\mathbf{S}_1}$ and $\mathbf{t}_{\mathbf{S}_2} \in R_{\mathbf{S}_2}$.

Using the above operations, *adaptive-consistency* is presented as in Figure 3.2. In step 1 the algorithm partitions the constraints into buckets whose structure depends on the variable ordering used. The main bucket operation is given in steps 4 and 5.

Algorithm adaptive-consistency specifies that it returns a "backtrack-free" network along the ordering d. This concept is related to the search approach that is common for solving constraint satisfaction, and in particular, to backtracking search (for more see Chapter 6). Backtracking search assign values to the variables in a certain order in a depth-first manner, checking the relevant constraints, until an assignment is made to all the variables or a *dead-end* is reached where no consistent values exist. If a dead-end is reached during search, the algorithm will *backtrack* to a previous variable, change its value, and proceed again along the ordering. We say that a constraint

ADAPTIVE-CONSISTENCY (AC)

Input: A constraint network $\mathcal{R} = \langle \mathbf{X}, \mathbf{D}, \mathbf{C} \rangle$, an ordering $d = (X_1, \ldots, X_n)$
Output: A backtrack-free network, denoted $E_d(\mathcal{R})$, along d, if the empty constraint was not generated. Else, the problem is inconsistent.

1. Partition constraints into $bucket_1$, ..., $bucket_n$ as follows:
   for $i \leftarrow n$ **downto** 1, put in $bucket_i$ all unplaced constraints mentioning X_i.
2. **for** $p \leftarrow n$ **downto** 1 **do**
3. **for** all the constraints $R_{\mathbf{S}_1}, \ldots, R_{\mathbf{S}_j}$ in $bucket_p$ **do**
4. $\mathbf{A} \leftarrow \bigcup_{i=1}^{j} \mathbf{S}_i - \{x_p\}$
5. $R_A \leftarrow \Pi_{\mathbf{A}}(\bowtie_{i=1}^{j} R_{\mathbf{S}_i})$
6. **if** $R_{\mathbf{A}}$ is not the empty relation **then** add $R_{\mathbf{A}}$ to the bucket of the latest variable in scope $\mathbf{A}$,
7. **else** exit and return the empty network
8. **return** $E_d(\mathcal{R}) = (\mathbf{X}, \mathbf{D}, bucket_1 \cup bucket_2 \cup \cdots \cup bucket_n)$

Figure 3.2: Adaptive-consistency as a bucket-elimination algorithm.

network is *backtrack-free* along an ordering d of its variables if it is guaranteed that a dead-end will never be encountered by backtracking search.

We next formally define the notion of backtrack-free network. It is based on the notion of a *partial solution*.

Definition 3.2 Partial solution. Given a constraint network $\mathcal{R}$, we say that an assignment of values to a subset of the variables $\mathbf{S} = \{X_1, ..., X_j\}$ denoted by $\mathbf{x}_1^j = (< X_1, x_1 >, < X_2, x_2 >, ..., < X_j, x_j >)$ is consistent relative to $\mathcal{R}$ iff it satisfies every constraint whose scope is subsumed in S. The assignment $\mathbf{x}_1^j$ (also called configuration) is also called a partial solution of $\mathcal{R}$.

Definition 3.3 Backtrack-free search. A constraint network is backtrack-free relative to a given ordering $d = (X_1, ..., X_n)$ if for every $j \leq n$, every partial solution $\mathbf{x}_1^j$ can be consistently extended to include X_{j+1}. Namely, there exists x_{j+1} s.t $\mathbf{x}_1^{j+1} = (\mathbf{x}_1^j, x_{j+1})$ is consistent.

We are now ready to state the main property of adaptive-consistency.

Theorem 3.4 Correctness and completeness of adapative-consistency. *[Dechter and Pearl, 1987] Given a set of constraints and an ordering of variables, adaptive-consistency decides if a set of* *constraints is consistent and, if it is, the algorithm always generates an equivalent representation that is backtrack-free along the input variable ordering. (Prove as an exercise.)*□

Example 3.5 Consider the graph coloring problem depicted in Figure 3.3 (modified by extending the domain of variable C with one additional color from Example 3.1 where colors are replaced by numbers). The figure shows a schematic execution of adaptive-consistency using the bucket data structure for the two orderings $d_1 = (E, B, C, D, A)$ and $d_2 = (A, B, D, C, E)$. The initial constraints, partitioned into buckets for both orderings, are displayed in the figure to the left of the double bars, while the constraints generated by the algorithm are displayed to the right of the double bar, in their respective buckets.

Notice that adaptive-consistency applied along ordering d_1 generates a different set of constraints, and in particular it generates only binary constraints (i.e., defined on pairs of variables), while along ordering d_2 the algorithm generates a ternary constraint. Notice also that for the ordering d_1, the constraint $B \neq E$ generated in $bucket_D$ is displayed for illustration only in $bucket_B$ (in parentheses), since there is already an identical original constraint. Indeed, the constraint is redundant.

Example 3.6 An alternative and more detailed graphical illustration of the algorithm's performance using d_2 is given in Figure 3.4. The figure shows, through the changing graph, how constraints are generated in the reverse order of $d_2 = A, B, D, C, E$ and how a solution is created in the forward order of d_2. The first step is processing the constraints that mention variables E. These are all joined to create a relation over $EDBC$ and then E is projected out, yielding a constraint on DBC whose relation is explicitly given in the figure. The relation is added to the set of constraints and is depicted as an added clique over the three nodes. Then E is removed, yielding the 3rd graph that includes only nodes A, D, B, C. The next variable to be processed is C. the constraints that include C are the original constraint $B \neq C$ and the new constraint over DBC. Joining both yields the new constraint on DBC as depicted, and projecting out C from this relation yields a constraint on DB whose meaning is the equality constraint. Variable D is eliminated next and then B, yielding the last variable A with two values $\{1, 2\}$.

Subsequently, the reverse process of generating a solution starts at A. Since it has two legitimate values, we can select any of those. The value $A = 1$ is selected. The next value satisfying the inequality constraint is $B = 2$, then $D = 2$ (satisfying $D = B$), then $C = 3$ (satisfying $C \neq B$). To assign a value for E we look at the constraint on $EDBC$ which only allows $E = 1$ to extend the current partial solution $A = 1, B = 2, D = 2, C = 3$, yielding a full solution.

What is the complexity of adaptive-consistency? It is clearly linear in the number of buckets and the time to process each bucket. However, since processing a bucket amounts to solving a constraint-satisfaction subproblem (generating the join of all relations) its complexity is exponential in the number of variables mentioned in a bucket. Conveniently, the number of variables

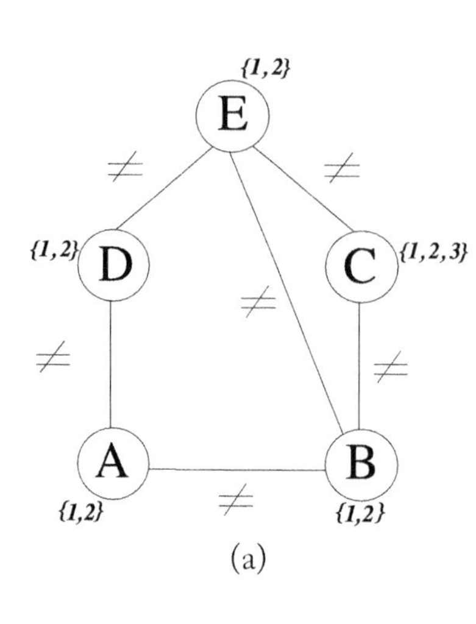

Ordering d_1

$Bucket(A)$: $A \neq D,\ A \neq B$
$Bucket(D)$: $D \neq E \ ||\ R_{DB}$
$Bucket(C)$: $C \neq B,\ C \neq E$
$Bucket(B)$: $B \neq E \ ||\ R^1_{BE}, R^2_{BE}$
$Bucket(E)$: $||\ R_E$

Ordering d_2
$Bucket(E)$: $E \neq D,\ E \neq C,\ E \neq B$
$Bucket(C)$: $C \neq B \ ||\ R_{DCB}$
$Bucket(D)$: $D \neq A \ ||\ R_{DB}(= D = B)$
$Bucket(B)$: $B \neq A \ ||\ R_{AB}(= R \neq B)$
$Bucket(A)$: $||\ R_A$

Figure 3.3: A modified graph coloring problem.

appearing in a bucket along a given ordering, can be obtained using the *induced-width* of the graph along that ordering. The induced-width is an important graph parameter that is instrumental to all bucket-elimination algorithms, and we define it next.

Definition 3.7 Induced-graph, width, and induced-width. Given an undirected graph $G = (V, E)$, where $V = \{v_1, ..., v_n\}$ is the set of nodes and E is a set of arcs over V. An *ordered graph* is a pair (G, d), where $d = (v_1, ..., v_n)$ is an ordering of the nodes. The nodes adjacent to v that precede it in the ordering are called its *parents*. The *width of a node* in an ordered graph is its number of parents. The *width of an ordered graph* (G, d), denoted $w(d)$, is the maximum width over all nodes. The *width of a graph* is the minimum width over all the orderings of the graph. The *induced-graph* of an ordered graph (G, d) is an ordered graph (G^*, d) where G^* is obtained from G as follows: the nodes of G are processed from last to first (top to bottom) along d. When a node v is processed, all of its parents are connected. The *induced width of an ordered graph*, (G, d), denoted $w^*(d)$, is the maximum number of parents a node has in the induced ordered graph (G^*, d). The *induced-width of a graph*, w^*, is the minimal induced width over all its orderings.

Example 3.8 Generating the induced-graph for $d_1 = E, B, C, D, A$ and $d_2 = A, B, D, C, E$ leads to the two graphs in Figure 3.5. The broken lines are the newly added arcs. The induced width along d_1 and d_2 are 2 and 3, respectively. They suggest different performance bounds for adaptive-consistency because the number of variables in a bucket is bounded by the number of parents of the corresponding variable plus 1 in the induced ordered graph which is equal to its induced-width plus 1.

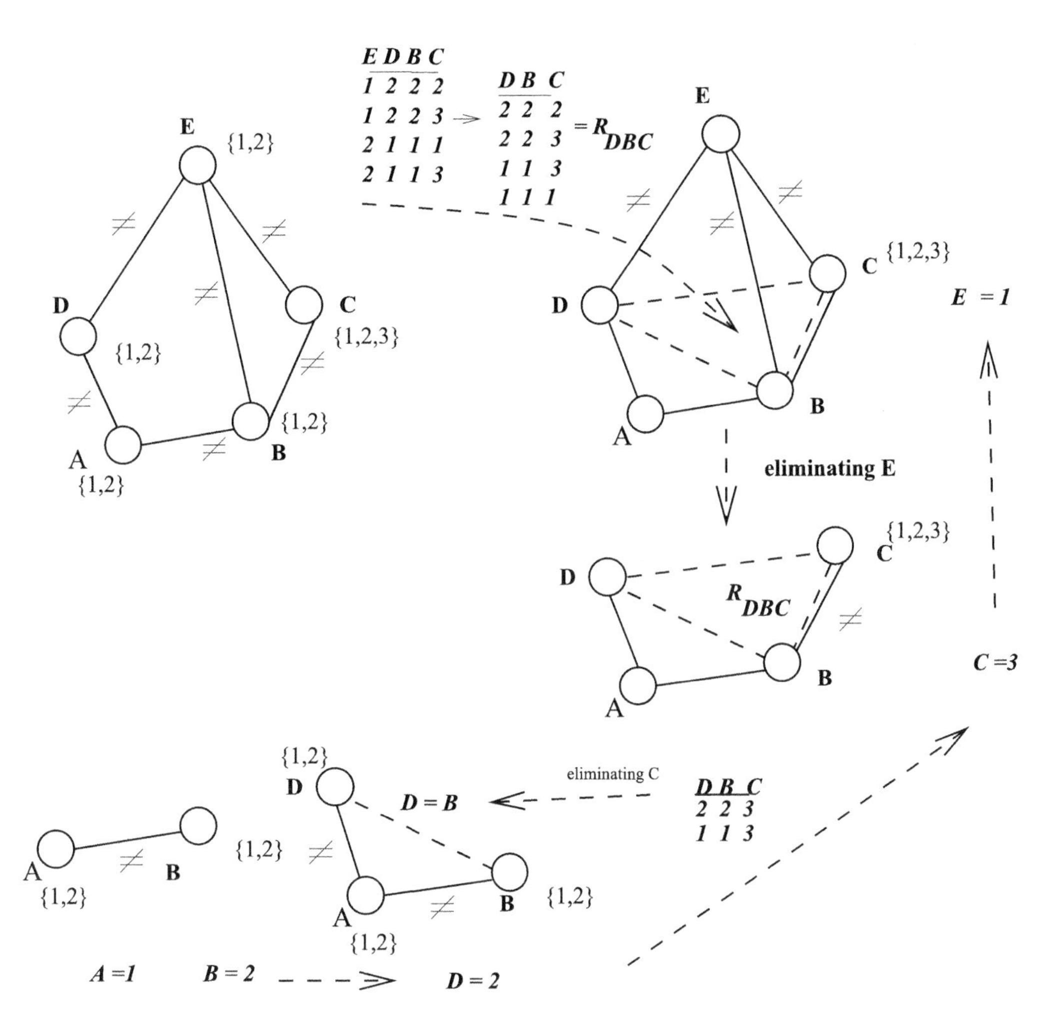

Figure 3.4: A schematic variable-elimination and solution-generation process is backtrack-free.

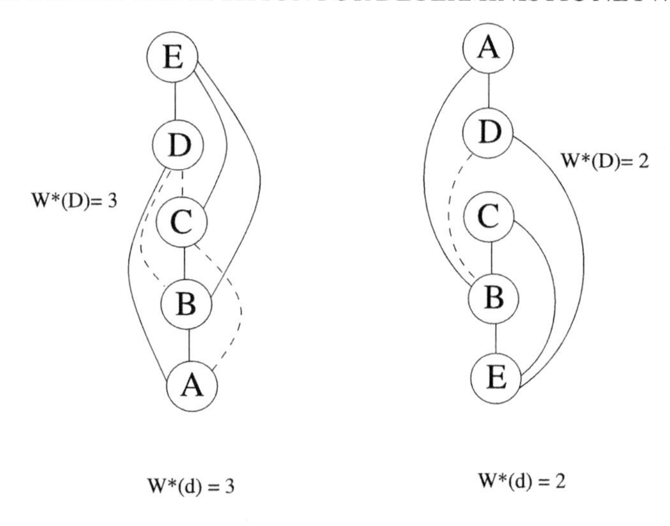

Figure 3.5: The induced width along the orderings: $d_1 = A, B, C, D, E$ and $d_2 = E, B, C, D, A$.

Theorem 3.9 *The time and space complexity of* ADAPTIVE-CONSISTENCY *is* $O((r+n)k^{w^*(d)+1})$ *and* $O(n \cdot k^{w^*(d)})$*, respectively, where* n *is the number of variables,* k *is the maximum domain size, and* $w^*(d)$ *is the induced-width along the order of processing* d *and* r *is the number of the problems' constraints.*

Proof: Since the total number of input functions plus those generated is bounded by $r+n$ and since the computation in a bucket is $O(r_i \cdot k^{w^*(d)+1})$, where r_i is the number of functions in $bucket_i$, a simple algebraic computation yields a total of $O((r+n)k^{w^*(d)+1})$. □

The above analysis suggests that problems having bounded induced width $w^* \leq b$ for some constant b can be solved in polynomial time. In particular, observe that when the graph is cycle-free its width and induced width are 1. Consider, for example, ordering $d = (A, B, C, D, E, F, G)$ for the tree in Figure 3.6. As demonstrated by the schematic execution along d, adaptive-consistency generates only unary relationships in this cycle-free graph.We note that on trees the algorithm can be accomplished in a distributed manner as a one-pass message passing algorithm.

3.2 BUCKET ELIMINATION FOR PROPOSITIONAL CNFS

Since propositional CNF formulas, discussed in Chapter 2, are a special case of constraint networks, we might wonder what adaptive consistency looks like when applied to them.

Propositional variables take only two values $\{true, false\}$ or "1" and "0." We denote propositional *variables* by uppercase letters $P, Q, R, \ldots$, propositional literals (i.e., $P =$"true" or $P =$"false") by P and $\neg P$ and disjunctions of literals, or *clauses*, are denoted by $\alpha, \beta, \ldots$. A *unit*

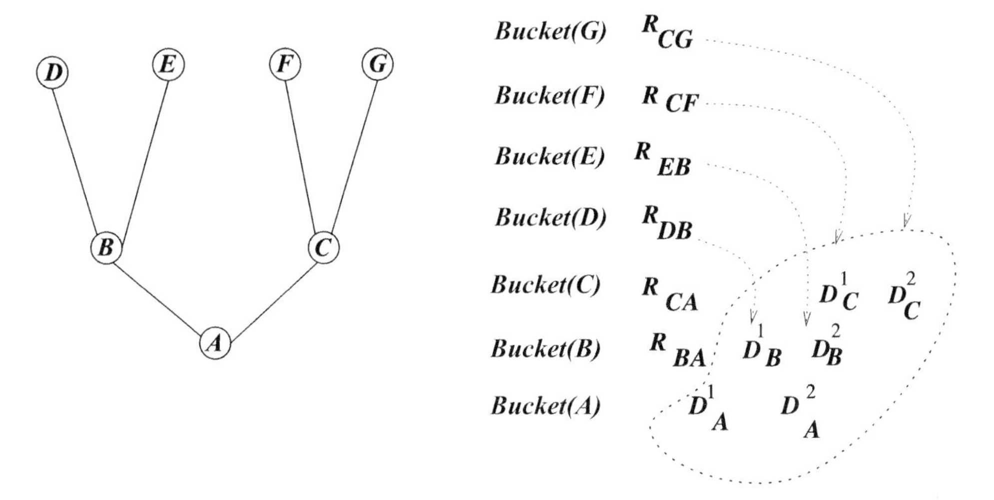

Figure 3.6: Schematic execution of adaptive-consistency on a tree network. D_X denotes unary constraints over X.

clause is a clause of size 1. The notation $(\alpha \vee T)$, when $\alpha = (P \vee Q \vee R)$ is shorthand for the disjunction $(P \vee Q \vee R \vee T)$. $(\alpha \vee \beta)$ denotes the clause whose literal appears in either α or β. The *resolution* operation over two clauses $(\alpha \vee Q)$ and $(\beta \vee \neg Q)$ results in a clause $(\alpha \vee \beta)$, thus eliminating Q. A formula φ in conjunctive normal form (*CNF*) is a set of clauses $\varphi = \{\alpha_1, \ldots, \alpha_t\}$ that denotes their conjunction. The set of *models* or *solutions* of a formula φ is the set of all truth assignments to all its symbols that do not violate any clause in φ. Deciding if a formula is satisfiable is known to be NP-complete [Garey and Johnson, 1979].

It turns out that the join-project operation used to process and eliminate a variable by adaptive-consistency over relational constraints translates to pair-wise resolution when applied to clauses [Rish and Dechter, 2000].

Definition 3.10 Extended composition. The extended composition of relation $R_{S_1}, \ldots, R_{S_m}$ relative to a subset of variables $\mathbf{A} \subseteq \bigcup_{i=1}^{m} \mathbf{S}_i$, denoted $EC_{\mathbf{A}}(R_{\mathbf{S}_1}, \ldots, R_{\mathbf{S}_m})$, is defined by

$$EC_{\mathbf{A}}(R_{\mathbf{S}_1}, \ldots, R_{\mathbf{S}_m}) = \pi_{\mathbf{A}}(\bowtie_{i=1}^{m} R_{\mathbf{S}_i}) .$$

When extended composition is applied to m relations, it is called *extended m-composition*. If the projection operation is restricted to subsets of size i, it is called *extended (i, m)-composition*.

It is not hard to see that extended composition is the operation applied in each bucket by adaptive-consistency. We next show that the notion of resolution is equivalent to extended 2-composition.

Lemma 3.11 *The* resolution *operation over two clauses,* $(\alpha \vee Q)$ *and* $(\beta \vee \neg Q)$*, results in a clause* $(\alpha \vee \beta)$ *for which* $models(\alpha \vee \beta) = EC_{scope(\alpha \vee \beta)}(models(\alpha \vee Q), models(\beta \vee \neg Q))$*. (Prove as an exercise.)* □

Example 3.12 Consider the two clauses $\alpha = (P \vee \neg Q \vee \neg O)$ and $\beta = (Q \vee \neg W)$. Now let the relation $R_{PQO} = \{000, 100, 010, 001, 110, 101, 111\}$ be the models of α and the relation $R_{QW} = \{00, 10, 11\}$ be the models of β. Resolving these two clauses over Q generates the resolvent clause $\gamma = res(\alpha, \beta) = (P \vee \neg O \vee \neg W)$. The models of γ are $\{(000, 100, 010, 001, 110, 101, 111\}$. The reader should verify that $EC_{PQW}(R_{PQO}, R_{QW})$ which is equal, by definition to $\pi_{RQW}(R_{PQO} \bowtie R_{Qw})$ yields the models of γ.

Substituting extended decomposition by resolution in adaptive consistency yields a bucket-elimination algorithm for propositional satisfiability which we call *directional resolution (DR)* given in Figure 3.7. We call the output theory (i.e., formula) of directional resolution, denoted $E_d(\varphi)$, the *directional extension* of φ. The following description of the algorithm may be familiar. Given an ordering $d = (Q_1, ..., Q_n)$, all the clauses containing Q_i that do not contain any symbol higher in the ordering are placed in the bucket of Q_i, denoted $bucket_i$. The algorithm processes the buckets in the reverse order of d. Processing of $bucket_i$ means *resolving* over Q_i all the possible pairs of clauses in the bucket and inserting the resolvents into appropriate lower buckets. Note that if the bucket contains a unit clause (Q_i or $\neg Q_i$), only unit resolutions are performed, namely resolutions involving a unit clause.

Example 3.13 Given the input theory $\varphi_1 = \{(\neg C), (A \vee B \vee C), (\neg A \vee B \vee E), (\neg B \vee C \vee D)\}$, and an ordering $d = (E, D, C, B, A)$, the theory is partitioned into buckets and processed by directional resolution in reverse order. Resolving over variable A produces a new clause $(B \vee C \vee E)$, which is placed in $bucket_B$. Resolving over B then produces clause $(C \vee D \vee E)$, which is placed in $bucket_C$. Finally, resolving over C produces clause $(D \vee E)$, which is placed in $bucket_D$. Directional resolution now terminates, since no resolution can be performed in $bucket_D$ and $bucket_E$. The output is a non-empty directional extension $E_d(\varphi_1)$. Once the directional extension is available, model generation can begin. There are no clauses in the bucket of E, the first variable in the ordering, and therefore E can also be assigned any value (e.g., $E = 0$). Given $E = 0$, the clause $(D \vee E)$ in $bucket_D$ implies $D = 1$, clause $\neg C$ in $bucket_C$ implies $C = 0$, and clause $(B \vee C \vee E)$ in $bucket_B$, together with the current assignments to C and E, implies $B = 1$. Finally, A can be assigned any value since both clauses in its bucket are satisfied by previous assignments. The initial partitioning into buckets along the ordering d as well as the buckets' contents generated by the algorithm are depicted in Figure 3.8.

DIRECTIONAL-RESOLUTION (DR)

Input: A *CNF* theory φ, an ordering $d = Q_1, \ldots, Q_n$ of its variables.

Output: A decision of whether φ is satisfiable. If it is, a theory $E_d(\varphi)$, equivalent to φ, else an empty directional extension.

1. **Initialize:** Generate an ordered partition of clauses into buckets $bucket_1$, ..., $bucket_n$, where $bucket_i$ contains all clauses whose highest variable is Q_i.
2. **for** $i \leftarrow n$ **downto** 1 process $bucket_i$:
3.   **if** there is a unit clause **then** (the instantiation step)
           apply unit-resolution in $bucket_i$ and place the resolvents in their right buckets.
           if the empty clause was generated, theory is not satisfiable.
4.   **else** resolve each pair $\{(\alpha \vee Q_i), (\beta \vee \neg Q_i)\} \subseteq bucket_i$.
           if $\gamma = \alpha \vee \beta$ is empty, return $E_d(\varphi) = \{\}$, the theory is not satisfiable
           else add γ it to the appropriate bucket of its highest proposition.
5. **return** $E_d(\varphi) \leftarrow \bigcup_i bucket_i$.

Figure 3.7: Directional-resolution.

As already observed in example 3.13, once all the buckets are processed, and if the empty clause was not generated, a truth assignment (model) can be assembled in a backtrack-free manner by consulting $E_d(\varphi)$, using the order d. We can show, indeed, that *DR* is guaranteed to generate a backtrack-free representation along the order of processing. This can be shown directly (prove as an exercise) or indirectly by proving that for bi-valued domains 2-composition is equivalent to full extended-composition, and then apply Theorem 3.4.

Theorem 3.14 Backtrack-free by DR. *Given a theory φ and an ordering of its variables d, the directional extension $E_d(\varphi)$ generated by DR is backtrack-free along d.* □

Not surprisingly, the complexity of directional-resolution is exponentially bounded (time and space) in the *induced width* of the theory's interaction graph along the order of processing. Notice that the graph of theory φ_1 along the ordering d (depicted also in Figure 3.8b) has an induced width of 3.

Lemma 3.15 *Given a theory φ and an ordering $d = (Q_1, \ldots, Q_n)$, if Q_i has at most w parents in the induced graph along d, then the bucket of Q_i in the output $E_d(\varphi)$ contains no more than 3^{w+1} clauses.*

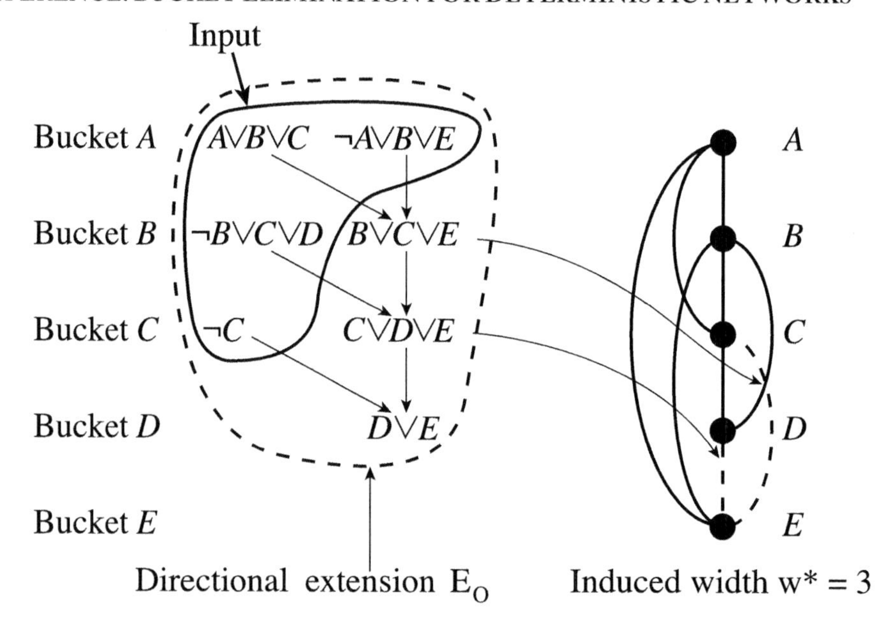

Figure 3.8: A schematic execution of directional resolution using ordering $d = (E, D, C, B, A)$.

Proof. Given a clause α in the bucket of Q_i, there are three possibilities for each parent P of Q_i: either P appears in α, $\neg P$ appears in α, or neither of them appears in α. Since Q_i also appears in α, either positively or negatively, the number of possible clauses in a bucket is no more than $2 \cdot 3^w < 3^{w+1}$. □

Since the number of parents of each variable is bounded by the induced-width along the order of processing we get the following.

Theorem 3.16 Complexity of DR. *Given a theory φ and an ordering of its variables d, the time complexity of algorithm DR along d is $O(n \cdot 9^{w^*(d)})$, and $E_d(\varphi)$ contains at most $n \cdot 3^{w^*(d+1)}$ clauses, where $w^*(d)$ is the induced width of φ's interaction graph along d.* □

3.3 BUCKET ELIMINATION FOR LINEAR INEQUALITIES

A special type of constraint is one that can be expressed by linear inequalities. The domains may be the real numbers, the rationals or finite subsets. In general, a linear constraint between r variables or less is of the form $\sum_{i=1}^{r} a_i X_i \leq c$, where a_i and c are rational constants. For example, $(3X_i + 2X_j \leq 3) \wedge (-4X_i + 5X_j \leq 1)$ are allowed constraints between variables X_i and X_j. In this special case, variable elimination amounts to the standard Gaussian elimination. From the inequalities $X - Y \leq 5$ and $X > 3$ we can deduce by eliminating X that $Y > 2$. The elimination operation is defined by the following.

Definition 3.17 Linear elimination. Let $\alpha = \sum_{i=1}^{(r-1)} a_i X_i + a_r X_r \leq c$, and $\beta = \sum_{i=1}^{(r-1)} b_i X_i + b_r X_r \leq d$. Then $elim_r(\alpha, \beta)$ is applicable only if a_r and b_r have opposite signs, in which case $elim_r(\alpha, \beta) = \sum_{i=1}^{r-1}(-a_i \frac{b_r}{a_r} + b_i) X_i \leq -\frac{b_r}{a_r} c + d$. If a_r and b_r have the same sign the elimination implicitly generates the universal constraint.

It is possible to show that the pair-wise join-project operation applied in a bucket can be accomplished by *linear elimination* as defined above. Applying adaptive-consistency to linear constraints and processing each pair of relevant inequalities in a bucket by linear elimination yields a bucket-elimination algorithm *Directional Linear Elimination* (abbreviated DLE), which is the well-known Fourier elimination algorithm (see [Lassez and Mahler, 1992]). For more information, see Chapter 8 in [Dechter, 2003].

Just one noteworthy comment regarding the bucket-elimination DLE (or Fourier elimination) is that it distinct itself from the general bucket-elimination class in its complexity. The complexity of Fourier elimination is not bounded exponentially by the induced width. The reason is that the number of linear inequalities that can be specified over a scope of size i cannot be bounded exponentially by i.

3.4 THE INDUCED-GRAPH AND INDUCED-WIDTH

We have seen that there is a tight relationship between the complexity of inference algorithms such as adaptive-consistency and the graph's induced-width. This algorithm, and all other inference algorithms that we will see are time and space exponential in the induced-width along the order of processing. This motivates finding an ordering with a smallest induced width, a task known to be hard [Arnborg, 1985]. However, useful greedy heuristics algorithms are available in the literature [Becker and Geiger, 1996; Dechter, 2003; Shoiket and Geiger, 1997]. To illustrate we use the following example.

Example 3.18 Consider Figure 3.9(a). For each ordering d, (G, d) is the graph depicted without the broken edges, while (G^*, d) is the corresponding induced graph that includes the broken edges. We see that the induced width of B along d_1 is 3, and that the overall induced width of this ordered graph is 3. The induced widths of the graph along orderings d_2 and d_3 both remain 2, and, therefore, the induced width of the graph G is 2.

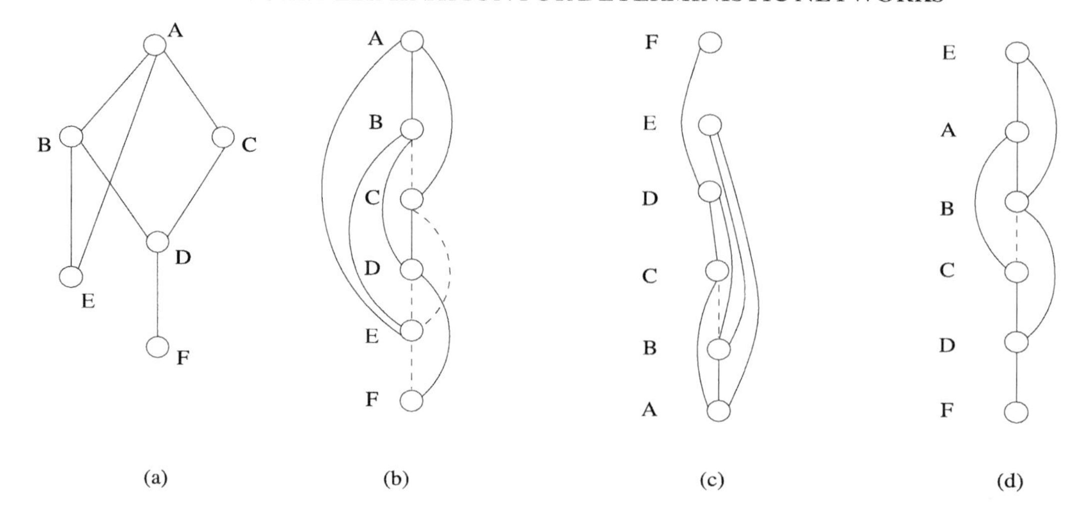

Figure 3.9: (a) Graph G, and three orderings of the graph; (b) $d_1 = (F, E, D, C, B, A)$, (c) $d_2 = (A, B, C, D, E, F)$, and (d) $d_3 = (F, D, C, B, A, E)$. Broken lines indicate edges added in the induced graph of each ordering.

3.4.1 TREES

A rather important observation is that a graph is a tree (has no cycles) iff it has an ordering whose width is 1. The reason a width-1 graph cannot have a cycle is because, if it has a cycle, for any ordering, at least one node on that cycle would have two parents, thus contradicting presumption of having a width-1 ordering. And vice-versa: if a graph has no cycles, it can always be converted into a rooted directed tree by directing all edges away from a designated root node. In such a directed tree, every node has exactly one node pointing to it; its parent. Therefore, any ordering in which, according to the rooted tree, every parent node precedes its child nodes, has a width of 1. Notice that given an ordering having width of 1, its induced-ordered graph has no additional arcs, yielding an induced width of 1, as well. In summary,

Proposition 3.19 *A graph is a tree iff it has width and induced width of 1.* □

3.4.2 FINDING GOOD ORDERINGS

Finding a minimum-width ordering of a graph, can be accomplished by various greedy algorithms. The greedy algorithm *min-width* (see Figure 3.10). The algorithm orders variables from last to first as follows: in the first step, a variable with minimum degree is selected and placed last in the ordering. The variable and all its adjacent edges are then eliminated from the original graph, and selection of the next variable continues recursively with the remaining graph. Ordering d_2 of G in Figure 3.9c could have been generated by a min-width ordering.

MIN-WIDTH (MW)

input: a graph $G = (V, E)$, $V = \{v_1, ..., v_n\}$
output: A min-width ordering of the nodes $d = (v_1, ..., v_n)$.
1. **for** $j = n$ to 1 by -1 do
2. $r \leftarrow$ a node in V with smallest degree.
3. put r in position j and $G \leftarrow G - \{r\}$.
   (delete from V node r and from E all its adjacent edges)
4. **endfor**

Figure 3.10: The min-width (MW) ordering procedure

Proposition 3.20 *[Freuder, 1982] Algorithm min-width* (MW) *finds a minimum width ordering of a graph and its complexity is $O(|E|)$ when E are the edges in the graph. (Prove as an exercise.)* □

Though finding the min-width ordering of a graph is easy, finding the minimum *induced width* of a graph is hard (NP-complete [Arnborg, 1985]). Nevertheless, deciding whether there exists an ordering whose induced width is less than a constant k, takes $O(n^k)$ time [S. A. Arnborg and Proskourowski, 1987].

A decent greedy algorithm, obtained by a small modification to the min-width algorithm, is the *min-induced-width* (MIW) algorithm (Figure 3.11). It orders the variables from last to first according to the following procedure: the algorithm selects a variable with minimum degree and places it last in the ordering. The algorithm next connects the node's neighbors in the graph to each other, and only then removes the selected node and its adjacent edges from the graph, continuing recursively with the resulting graph. The ordered graph in Figure 3.9c could also have been generated by a min-induced-width ordering of G. In this case, it so happens that the algorithm achieves w^*, the minimum induced width of the graph.

Another variation yields a greedy algorithm known as *min-fill.* It uses the *min-fill set*, that is, the number of edges needed to be filled so that the node's parent set be fully connected, as an ordering criterion. This *min-fill* heuristic described in Figure 3.12, was demonstrated empirically to be somewhat superior to min-induced-width algorithm [Kjæaerulff, 1990]. The ordered graph in Figure 3.9c could also have been generated by a min-fill ordering of G while the ordering d_1 or d_3 in parts (a) and (d) could not.

What is the complexity of MIW and MIN-Fill? It is easy to see that their complexity is bounded by $O(n^3)$.

The notions of width and induced width and their relationships to various graph parameters, have been studied extensively. In the following we briefly note the connection with chordal graphs.

MIN-INDUCED-WIDTH (MIW)
input: a graph $G = (V, E)$, $V = \{v_1, ..., v_n\}$
output: An ordering of the nodes $d = (v_1, ..., v_n)$.
1. **for** $j = n$ to 1 by -1 do
2. $r \leftarrow$ a node in V with smallest degree.
3. put r in position j.
4. connect r's neighbors: $E \leftarrow E \cup \{(v_i, v_j) | (v_i, r) \in E, (v_j, r) \in E\}$,
5. remove r from the resulting graph: $V \leftarrow V - \{r\}$.

Figure 3.11: The min-induced-width (MIW) procedure

MIN-FILL (MIN-FILL)
input: a graph $G = (V, E)$, $V = \{v_1, ..., v_n\}$
output: An ordering of the nodes $d = (v_1, ..., v_n)$.
1. **for** $j = n$ to 1 by -1 do
2. $r \leftarrow$ a node in V with smallest fill edges for his parents.
3. put r in position j.
4. connect r's neighbors: $E \leftarrow E \cup \{(v_i, v_j) | (v_i, r) \in E, (v_j, r) \in E\}$,
5. remove r from the resulting graph: $V \leftarrow V - \{r\}$.

Figure 3.12: The min-fill (MIN-FILL) procedure

3.5 CHORDAL GRAPHS

For some special graphs such as chordal graphs, computing the induced-width is easy. A graph is *chordal* if every cycle of length at least four has a chord, that is, an edge connecting two nonadjacent vertices. For example, G in Figure 3.9a is not chordal since the cycle (A, B, D, C, A) does not have a chord. The graph can be made chordal if we add the edge (B, C) or the edge (A, D).

Many difficult graph problems become easy on chordal graphs. For example, finding all the maximal (largest) *cliques* (completely connected subgraphs) in a graph, an NP-complete task on general graphs, is easy for chordal graphs. This task (finding maximal cliques) is facilitated by using yet another ordering procedure called the *max-cardinality ordering* [Tarjan and Yannakakis, 1984]. A *max-cardinality ordering* of a graph orders the vertices from *first to last* according to the following rule: the first node is chosen arbitrarily. From this point on, a node that is connected to a maximal number of already ordered vertices is selected, and so on (see algorithm in Figure 3.13). Ordering d_2 in Figure 3.9c,d are max-cardinality ordering but ordering d_1 is not.

A max-cardinality ordering can be used to identify chordal graphs. Namely, a graph is chordal iff in a max-cardinality ordering each vertex and all its parents form a clique. One can thereby enumerate all maximal cliques associated with each vertex (by listing the sets of each vertex and its parents, and then identify the maximal size of a clique). Notice that there are at most n maximal cliques: each vertex and its parents is one such clique. In addition, when using a max-cardinality ordering of a chordal graph, the ordered graph is identical to its induced graph, and therefore its width is identical to its induced width. It is easy to see that,

Proposition 3.21 *If G is the induced graph of a graph G, along some ordering d, then G is chordal.* □

Proof. One way to show this is to realize that the ordering d can be obtained as a max-cardinality ordering of G. □

Example 3.22 We see again that G in Figure 3.9a is not chordal since the parents of A are not connected in the max-cardinality ordering in Figure 3.9d. If we connect B and C, the resulting induced graph is chordal.

MAX-CARDINALITY (MC)
input: a graph $G = (V, E)$, $V = \{v_1, ..., v_n\}$
output: An ordering of the nodes $d = (v_1, ..., v_n)$.
1. Place an arbitrary node in position 0.
2. **for** $j = 1$ to n do
3. $r \leftarrow$ a node in G that is connected to a largest subset of nodes in positions 1 to $j - 1$, breaking ties arbitrarily.
4. **endfor**

Figure 3.13: The max-cardinality (MC) ordering procedure

A very important attraction for max-cardinality ordering is that it can be achieved in linear time.

Proposition 3.23 *[Tarjan and Yannakakis, 1984] Given a graph $G = (V, E)$ the complexity of max-cardinality search is $O(n + m)$ when $|V| = n$ and $|E| = m$.*

A subclass of chordal graphs are *k-trees*. A *k-tree* is a chordal graph whose maximal cliques are of size $k + 1$, and it can be defined recursively as follows:

Definition 3.24 k-trees. A complete graph with k vertices is a *k-tree*. A *k-tree* with r vertices can be extended to $r + 1$ vertices by connecting the new vertex to all the vertices in any clique of size k. A partial k-tree is a k-tree having some of its arcs removed. Namely it will clique of size smaller than k.

3.6 SUMMARY AND BIBLIOGRAPHY NOTES

This chapter is based in part on Chapters 4 and 8 of [Dechter, 2003]. It introduces the inference bucket-elimination algorithm for constraint satisfaction problems, called ADAPTIVE-CONSISTENCY and shows the principled connection between this class of algorithms and the graph-parameter of *induced-width*. The algorithm complexity is exponentially bounded by the induced-width along the order of processing. Subsequently, the algorithm is applied to propositional formulas in conjunctive normal form for solving satisfiability. The connection between pair-wise resolution and the operation of variable elimination is described and the resulting *directional resolution* algorithm is shown to be a particular case of adaptive-consistency and therefore of bucket elimination as well. The connection to variable elimination over linear inequalities is also briefly noted, showing that Fourier elimination is also a special case of this class of algorithms even though it's complexity is not bounded exponentially by the induced-width. The chapter conclude with a brief discussion on relevant graph algorithms for orderings.

Algorithm ADAPTIVE-CONSISTENCY was introduced by Dechter [Dechter and Pearl, 1987] as well as its complexity analysis using the concept of *induced-width* as the principle graph-parameter that controls the algorithms's complexity. A similar elimination algorithm was introduced earlier by Seidel [Seidel, 1981]. It was observed that these algorithms belong to the class of Dynamic Programming algorithms as presented in [Bertele and Brioschi, 1972]. In [Dechter and Pearl, 1989], the connection between ADAPTIVE-CONSISTENCY and tree-clustering algorithms was made explicit, as will will show in Chapter 5.

The observation that pair-wise resolution is the variable-elimination operation for CNFs in ADAPTIVE-CONSISTENCY yielded algorithm DIRECTIONAL-RESOLUTION for CNFs, which was presented in [Dechter and Rish, 1994; Rish and Dechter, 2000]. It was also observed that the resulting algorithm is the well-known David-Putnam (DP) algorithm [Davis and Putnam, 1960] implying that the complexity of DP is exponentially by the induced-width as well.

CHAPTER 4

Inference: Bucket Elimination for Probabilistic Networks

Having investigated bucket elimination in deterministic constraint networks in the previous chapter, we now present the bucket-elimination algorithm for the three primary queries defined over probabilistic networks: (1) belief-updating or computing posterior marginals (bel) as well as finding the probability of evidence; (2) finding the most probable explanation (mpe); and (3) finding the maximum a posteriori hypothesis (map).

We start focusing on queries over Bayesian networks first, and later show that the algorithms we derive are applicable with minor changes to Markov networks, cost networks, and mixed networks as well to the general graphical model. Below we recall the definition of Bayesian networks for your convenience.

Definition 4.1 (Bayesian networks) A *Bayesian network (BN)* is a 4-tuple $\mathcal{B} = \langle \mathbf{X}, \mathbf{D}, \mathbf{P}_G, \prod \rangle$. $\mathbf{X} = \{X_1, \ldots, X_n\}$ is a set of ordered variables defined over domains $\mathbf{D} = \{D_1, \ldots, D_n\}$, where $d = (X_1, \ldots, X_n)$ is an ordering of the variables. The set of functions $P_G = \{P_1, \ldots, P_n\}$ consist of conditional probability tables (CPTs for short) $P_i = \{P(X_i | \mathbf{Y}_i)\}$ where $\mathbf{Y}_i \subseteq \{X_{i+1}, ..., X_n\}$. These P_i functions can be associated with a directed acyclic graph G in which each node represents a variable X_i and $\mathbf{Y}_i = pa(X_i)$ are the parents of X_i in G. That is, there is a directed arc from each parent variable of X_i to X_i. The Bayesian network $\mathcal{B}$ represents the probability distribution over $\mathbf{X}$, $P_{\mathcal{B}}(x_1, \ldots, x_n) = \prod_{i=1}^{n} P(x_i | \mathbf{x}_{pa(X_i)})$. We define an evidence set e as an instantiated subset of the variables.

4.1 BELIEF UPDATING AND PROBABILITY OF EVIDENCE

Belief updating is the primary inference task over Bayesian networks. The task is to determine the posterior probability of singleton variables once new evidence arrives. For instance, if we are interested in the likelihood that the sprinkler was on last night (as we were in the Bayesian network example in Chapter 2), then we need to update this likelihood if we observe that the pavement near the sprinkler is slippery. More generally, we are sometime asked to compute the posterior marginals of a subset of variables given some evidence. Another important query over Bayesian networks, computing the probability of the evidence, namely computing the joint likelihood of a

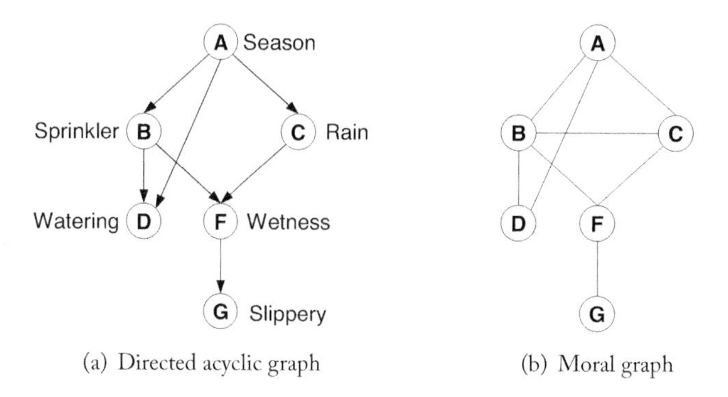

(a) Directed acyclic graph (b) Moral graph

Figure 4.1: Belief network $P(G, F, C, B, A) = P(G|F)P(F|C, B)P(D|A, B)P(C|A)P(B|A)P(A)$.

specific assignment to a subset of variables, is highly related to belief updating. We will show in this chapter how these tasks can be computed by the bucket-elimination scheme.

4.1.1 DERIVING BE-BEL

We next present a step-by-step derivation of a general variable-elimination algorithm for belief updating. This algorithm is similar to adaptive-consistency, but the join and project operators of adaptive-consistency are replaced, respectively, with the operations of product and summation. We begin with an example and then proceed to describe the general case.

Let $X_1 = x_1$ be an atomic proposition (e.g., pavement = slippery). The problem of belief updating is to compute the conditional probability of X_1 given evidence $\mathbf{e}$, $P(x_1|\mathbf{e})$, and the probability of the evidence $P(\mathbf{e})$. By Bayes rule we have that $P(x_1|\mathbf{e}) = \frac{P(x_1,\mathbf{e})}{P(\mathbf{e})}$, where $\frac{1}{P(\mathbf{e})}$ is called the normalization constant. To develop the algorithm, we will use the previous example of a Bayesian network, (Figure 2.5), and assume the evidence is $G = 1$ (we also use $g = 1$ meaning that the specific assignment g is the number 1). For convenience we depict here part of the network again in Figure 4.1.

Consider the variables in the ordering $d_1 = A, C, B, F, D, G$. We want to compute $P(A = a|g = 1)$ or $P(A = a, g = 1)$. By definition,

$$P(a, g = 1) = \sum_{c,b,e,d,g=1} P(a, b, c, d, e, g) = \sum_{c,b,f,d,g=1} P(g|f)P(f|b, c)P(d|a, b)P(c|a)P(b|a)P(a$$

We can now apply some simple symbolic manipulation, migrating each conditional probability table to the left of the summation variables that it does not reference. We get

$$P(a, g = 1) = P(a)\sum_{c} P(c|a)\sum_{b} P(b|a)\sum_{f} P(f|b, c)\sum_{d} P(d|b, a)\sum_{g=1} P(g|f). \quad (4.1)$$

Carrying the computation from right to left (from G to A), we first compute the rightmost summation, which generates a function over F that we denote by $\lambda_G(F)$, defined by:

$\lambda_G(f) = \sum_{g=1} P(g|f)$ and place it as far to the left as possible, yielding

$$P(a, g = 1) = P(a) \sum_c P(c|A) \sum_b P(b|a) \sum_f P(f|b, c)\lambda_G(f) \sum_d P(d|b, a). \tag{4.2}$$

(We index a generated function by the variable that was summed over to create it; for example, we created $\lambda_G(f)$ by summing over G.) Summation removes or eliminates a variable from the calculation.

Summing next over D (generating a function denoted $\lambda_D(B, A)$, defined by $\lambda_D(a, b) = \sum_d P(d|a, b)$), we get

$$P(a, g = 1) = P(a) \sum_c P(c|a) \sum_b P(b|a)\lambda_D(a, b) \sum_f P(f|b, c)\lambda_G(f) . \tag{4.3}$$

Next, summing over F (generating $\lambda_F(B, C)$ defined by $\lambda_F(b, c) = \sum_f P(f|b, c)\lambda_G(f)$), we get,

$$P(a, g = 1) = P(a) \sum_c P(c|a) \sum_b P(b|a)\lambda_D(a, b)\lambda_F(b, c) . \tag{4.4}$$

Summing over B (generating $\lambda_B(A, C)$), we get

$$P(a, g = 1) = P(a) \sum_c P(c|a)\lambda_B(a, c) . \tag{4.5}$$

Finally, summing over C (generating $\lambda_C(A)$), we get

$$P(a, g = 1) = P(a)\lambda_C(a) . \tag{4.6}$$

Summing over the values of variable A, we generate $P(g = 1) = \sum_a P(a)\lambda_C(a)$. The answer to the query $P(a|g = 1)$ can be computed by normalizing the last product in Eq. (4.6). Namely, $P(a|g = 1) = \alpha P(a)\lambda_C(a)$ where $\alpha = \frac{1}{P(g=1)}$.

We can create a bucket-elimination algorithm for this calculation by mimicking the above algebraic manipulation, using *buckets* as the organizational device for the various sums. First, we partition the $CPTs$ into buckets relative to the given order, $d_1 = A, C, B, F, D, G$. In bucket G we place all functions mentioning G. From the remaining $CPTs$ we place all those mentioning D in $bucket_D$, and so on. This is precisely the partition rule we used in the adaptive-consistency algorithm for constraint networks. This results in the initial partitioning given in Figure 4.2. Note that observed variables are also placed in their corresponding bucket.

Initializing the buckets corresponds to deriving the expression in Eq. (4.1). Now we process the buckets from last to first (or top to bottom in the figures), implementing the right to left computation in Eq. (4.1). Processing a bucket amounts to eliminating the variable in the bucket from subsequent computation. The $bucket_G$ is processed first. We eliminate G by summing over all values of G, but since we have observed that $G = 1$, the summation is over a singleton value.

$$\begin{aligned}
Bucket_G &= P(g|f), g = 1\\
Bucket_D &= P(d|b,a)\\
Bucket_F &= P(f|b,c)\\
Bucket_B &= P(b|a)\\
Bucket_C &= P(c|a)\\
Bucket_A &= P(a)
\end{aligned}$$

Figure 4.2: Initial partitioning into buckets using $d_1 = A, C, B, F, D, G$.

The function $\lambda_G(f) = \sum_{g=1} P(g|f) = P(g = 1|f)$, is computed and placed in $bucket_F$. In our calculations above, this corresponds to deriving Eq. (4.2) from Eq. (4.1)). Once we have created a new function, it is placed in a lower bucket in accordance with the same rule we used to partition the original $CPTs$.

Following order d_1, we proceed by processing $bucket_D$, summing over D the product of all the functions that are in its bucket. Since there is a single function, the resulting function is $\lambda_D(b,a) = \sum_d P(d|b,a)$ and it is placed in $bucket_B$. Subsequently, we process the buckets for variables F, B, and C in order, each time summing over the relevant variable and moving the generated function into a lower bucket according to the same placement rule. In $bucket_A$ we compute the answer $P(a|g = 1) = \alpha \cdot P(a) \cdot \lambda_C(a)$. Figure 4.3 summarizes the flow of functions generated during the computation.

In this example, the generated λ functions were at most two-dimensional; thus, the complexity of processing each bucket using ordering d_1 is (roughly) time and space quadratic in the domain sizes. But would this also be the case had we used a different variable ordering? Consider ordering $d_2 = A, F, D, C, B, G$. To enforce this ordering we require that the summations remain in order d_2 from right to left, yielding (and we leave it to you to figure how the λ functions are generated):

$$\begin{aligned}
&P(a, g = 1) = P(a) \textstyle\sum_f \sum_d \sum_c P(c|a) \sum_b P(b|a)\ P(d|a,b)P(f|b,c) \sum_{g=1} P(g|f)\\
&= P(a) \textstyle\sum_f \lambda_G(f) \sum_d \sum_c P(c|a) \sum_b P(b|a)\ P(d|a,b)P(f|b,c)\\
&= P(a) \textstyle\sum_f \lambda_G(f) \sum_d \sum_c P(c|a)\lambda_B(a,d,c,f)\\
&= P(a) \textstyle\sum_f \lambda_g(f) \sum_d \lambda_C(a,d,f)\\
&= P(a) \textstyle\sum_f \lambda_G(f)\lambda_D(a,f)\\
&= P(a)\lambda_F(a).
\end{aligned}$$

The analogous bucket-elimination schematic process for this ordering is shown in Figure 4.4a. As before, we finish by calculating $P(a|g = 1) = \alpha P(a)\lambda_F(a)$, where $\alpha = \frac{1}{\sum_a P(a)\lambda_F(a)}$.

We conclude this section with a general derivation of the bucket-elimination algorithm for probabilistic networks, called *BE-bel*. As a byproduct, the algorithm yields the probability of the evidence. Consider an ordering of the variables $d = (X_1, ..., X_n)$ and assume we seek $P(X_1|e)$.

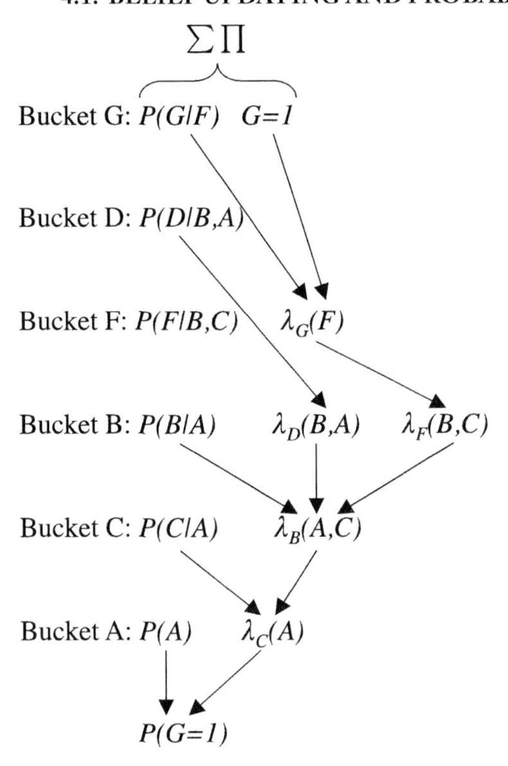

Figure 4.3: Bucket elimination along ordering $d_1 = A, C, B, F, D, G$.

Note that if we seek the belief for variable X_1 it should initiate the ordering. Later we will see how this requirement can be relaxed. We need the definition of ordered-restricted Markov blanket.

Definition 4.2 Markov blanket. Given a Bayesian network $\mathcal{B} = \langle \mathbf{X}, \mathbf{D}, \mathbf{P}_G, \prod \rangle$ and an ordering $d = (X_1, ..., X_n)$ the Markov blanket of variable X_j relative to d, denoted M_j is the set of its neighbors in the moral graph that precede it in the ordering d, including X_j. We denote by $S_j = M_j - \{X_j\}$.

Using the notation $\mathbf{x}_{(i..j)} = (x_i, x_{i+1}, ..., x_j)$ and $\mathbf{x} = (x_1, ..., x_n)$ we want to compute:

$$P(x_1, \mathbf{e}) = \sum_{\mathbf{x}_{(2..n)}} P(\mathbf{x}, \mathbf{e}) = \sum_{\mathbf{x}_{(2..n-1)}} \sum_{x_n} \prod_i P(x_i, \mathbf{e} | \mathbf{x}_{pa_i}) \, .$$

Separating X_n from the rest of the variables results in:

$$= \sum_{\mathbf{x}_{(2..n-1)}} \prod_{X_i \in X - M_n} P(x_i, \mathbf{e} | \mathbf{x}_{pa_i}) \cdot \sum_{x_n} \prod_{X_i \in M_n} P(x_i, \mathbf{e} | \mathbf{x}_{pa_i})$$

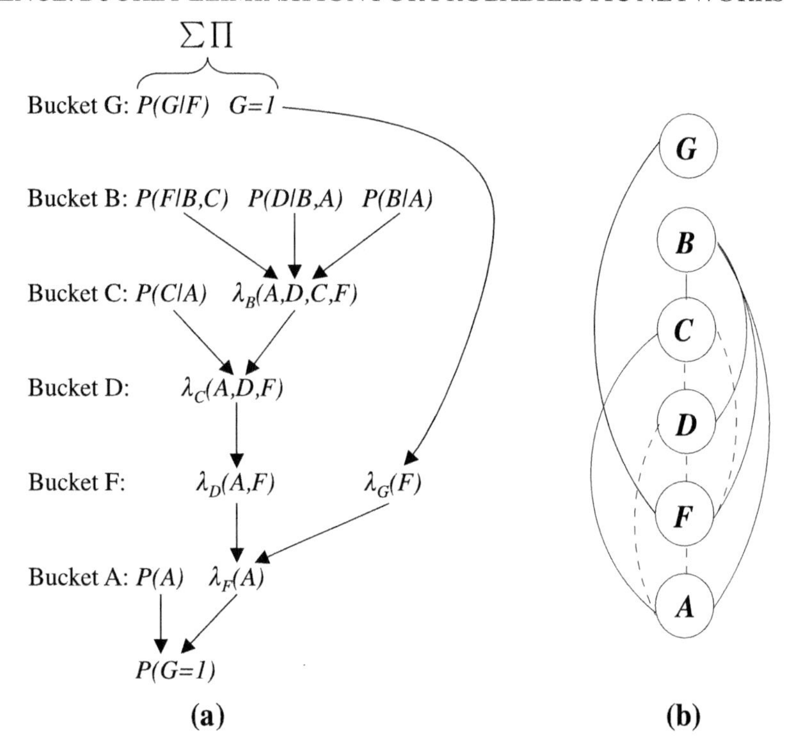

Figure 4.4: The bucket's output when processing along $d_2 = A, F, D, C, B, G$.

$$= \sum_{\mathbf{x}_{(2..n-1)}} \prod_{X_i \in X - M_n} P(x_i, \mathbf{e}|\mathbf{x}_{pa_i}) \cdot \lambda_n(\mathbf{x}_{S_n})$$

where

$$\lambda_n(\mathbf{x}_{S_n}) = \sum_{x_n} \prod_{X_i \in M_n} P(x_i, \mathbf{e}|\mathbf{x}_{pa_i}). \tag{4.7}$$

The process continues recursively with X_{n-1}.

Thus, the computation performed in bucket X_n is captured by Eq. (4.7). Given ordering $d = X_1, ..., X_n$, where the queried variable appears first, the CPTs are partitioned using the rule described earlier. The computed function in $bucket_{X_p}$ is λ_{X_p} and is defined over the bucket's scope, excluding X_p $\mathbf{S}_p = \cup_{\lambda_i \in bucket_{X_p}} scope(\lambda_i) - X_p$ by $\lambda_X = \sum_{X_p} \psi_X \cdot \prod_{\lambda \in bucket_{X_p}} \lambda$, where, $\psi_X = \prod_{P \in bucket_X} P$. This function is placed in the bucket of its largest-index variable in $\mathbf{S}_p$. Once processing reaches the first bucket, we have all the information to compute the answer which is the product of those functions. If we also *process* the first bucket we get the probability of the evidence. Algorithm BE-bel is described in Figure 4.5 (step 3 will elaborate more shortly). With the above derivation we showed the following.

ALGORITHM BE-BEL

Input: A belief network $\mathcal{B} = \langle \mathbf{X}, \mathbf{D}, \mathbf{P}_G, \prod \rangle$, an ordering $d = (X_1, \ldots, X_n)$; evidence e
output: The belief $P(X_1|\mathbf{e})$ and probability of evidence $P(\mathbf{e})$

1. Partition the input functions (CPTs) into $bucket_1$, ..., $bucket_n$ as follows: **for** $i \leftarrow n$ **downto** 1, put in $bucket_i$ all unplaced functions mentioning X_i. Put each observed variable in its bucket. Denote by ψ_i the product of input functions in $bucket_i$.
2. **backward: for** $p \leftarrow n$ **downto** 1 **do**
3. **for** all the functions $\psi_{S_0}, \lambda_{S_1}, \ldots, \lambda_{S_j}$ in $bucket_p$ **do**
   If (observed variable) $X_p = x_p$ appears in $bucket_p$,
   assign $X_p = x_p$ to each function in $bucket_p$ and then
   put each resulting function in the bucket of the *closest* variable in its scope.
   else,
4. $\lambda_p \leftarrow \sum_{X_p} \psi_p \cdot \prod_{i=1}^{j} \lambda_{S_i}$
5. place λ_p in bucket of the latest variable in scope(λ_p),
6. **return** (as a result of processing $bucket_1$):
   $P(\mathbf{e}) = \alpha = \sum_{X_1} \psi_1 \cdot \prod_{\lambda \in bucket_1} \lambda$
   $P(X_1|\mathbf{e}) = \frac{1}{\alpha}\psi_1 \cdot \prod_{\lambda \in bucket_1} \lambda$

Figure 4.5: BE-bel: a sum-product bucket-elimination algorithm.

Theorem 4.3 Sound and complete. *Algorithm BE-Bel applied along any ordering that starts with X_1 computes the belief $P(X_1|\mathbf{e})$. It also computes the probability of evidence $P(\mathbf{e})$ as the inverse of the normalizing constant in the first bucket.* □

The bucket's operations for BE-bel

Processing a bucket requires the two types of operations on the functions in the buckets, combinations, and marginalization. The combination operation in this case is a product, which generates a function whose scope is the union of the scopes of the bucket's functions. The marginalization operation is summation, summing out the bucket's variable. The algorithm often referred to as being a *sum-product algorithm*.

Example 4.4 Let's look at an example of both of these operations in a potential bucket of B assuming it contains only two functions, $P(F|B,C)$ and $P(B|A)$. These functions are displayed in Figure 4.6. To take the product of the functions $P(F|B,C)$ and $P(B|A)$ we create a function over F, B, C, A where for each tuple assignment, the function value is the product of the respective entries in the input functions. To eliminate variable B by summation, we sum the function generated by the product, over all values in of variable B. We say that we *sum out* variable B. The computation of both the product and summation operators are depicted in Figure 4.7.

B	C	F	$P(F\|B,C)$
false	false	true	0.1
true	false	true	0.9
false	true	true	0.8
true	true	true	0.95

A	B	$P(B\|A)$
summer	false	0.2
fall	false	0.6
winter	false	0.9
spring	false	0.4

Figure 4.6: Examples of functions in the bucket of B.

The implementation details of the algorithm to perform these operations might have a significant impact on the performance. In particular, much depends on how the bucket's functions are represented. If, for example, the $CPTs$ are represented as matrices, then we can exploit efficient matrix multiplication algorithms. This important issue is outside the scope of this book.

4.1.2 COMPLEXITY OF BE-bel

Although BE-bel can be applied along any ordering, its complexity varies considerably across different orderings. Using ordering d_1 we recorded λ functions on pairs of variables only, while using d_2 we had to record functions on as many as four variables (see $Bucket_C$ in Figure 4.4a). The arity (i.e., the scope size) of the function generated during processing of a bucket equals the number of variables appearing in that processed bucket, excluding the bucket's variable itself. Since computing and recording a function of arity r is time and space exponential in r we conclude that the complexity of the algorithm is dominated by its largest scope bucket and it is therefore exponential in the size (number of variables) of the bucket having the largest number of variables. The base of the exponent is bounded by a variable's domain size.

Fortunately, as was observed earlier for adaptive-consistency, the bucket sizes can be easily predicted from the elimination process along the ordered graph. Consider the *moral graph* of a given Bayesian network. This graph has a node for each variable and any two variables appearing in the same CPT are connected. The moral graph of the network in Figure 4.1(a) is given in Figure

A	B	C	F	$f(A, B, C, F) = P(F\|B, C) \cdot P(B\|A)$
summer	false	false	true	0.2 × 0.1 =0.02
summer	false	true	true	0.2 × 0.8 =0.16
fall	false	false	true	0.6 × 0.1 = 0.06
fall	false	true	true	0.6 × 0.8 = 0.46
winter	false	false	true	0.9 × 0.1 = 0.09
winter	false	true	true	0.9 × 0.8 = 0.72
spring	false	false	true	0.4 × 0.1 =0.04
spring	false	true	true	0.4 × 0.8 =0.32
summer	true	false	true	0.8 × 0.9 =0.72
summer	true	true	true	0.8 × 0.95 =0.76
fall	true	false	true	0.4 × 0.9 =0.36
fall	true	true	true	0.4 × 0.95 =0.38
winter	true	false	true	0.1 × 0.9 =0.09
winter	true	true	true	0.1 × 0.95 =0.095
spring	true	false	true	0.6 × 0.9 =0.42
spring	true	true	true	0.6 × 0.95 =0.57

A	C	F	$\lambda_B(A, B, F) = \sum_B f(A, B, C, F)$
summer	false	true	0.02+0.72 = 0.74
fall	false	true	0.06 + 0.36 = 0.42
winter	false	true	0.09 +0.09 = 0.18
spring	false	true	0.04 + 0.42 = 0.46
summer	true	true	0.72 + 0.16 = 0.88
fall	true	true	0.46 + 0.38 = 0.84
winter	true	true	0.72 + 0.095 = 0.815
spring	true	true	0.32 + 0.57 = 0.89

Figure 4.7: Processing the functions in the bucket of B.

4.1(b). If we take this moral graph and impose an ordering on its nodes, the induced-width of the ordered graph of each nodes captures the number of variables which would be processed in that bucket. We demonstrate this next.

Example 4.5 Recall the definition of induced graph (Definition 3.7). The induced moral graph in Figure 4.8, relative to ordering $d_1 = A, C, B, F, D, G$ is depicted in Figure 4.8a. Along this ordering the induced ordered graph was not added any edges over the original graph, since all the earlier neighbors of each node are already connected. The induced width of this graph is 2. Indeed, in this case, the maximum arity of functions recorded by the algorithm is 2. For ordering

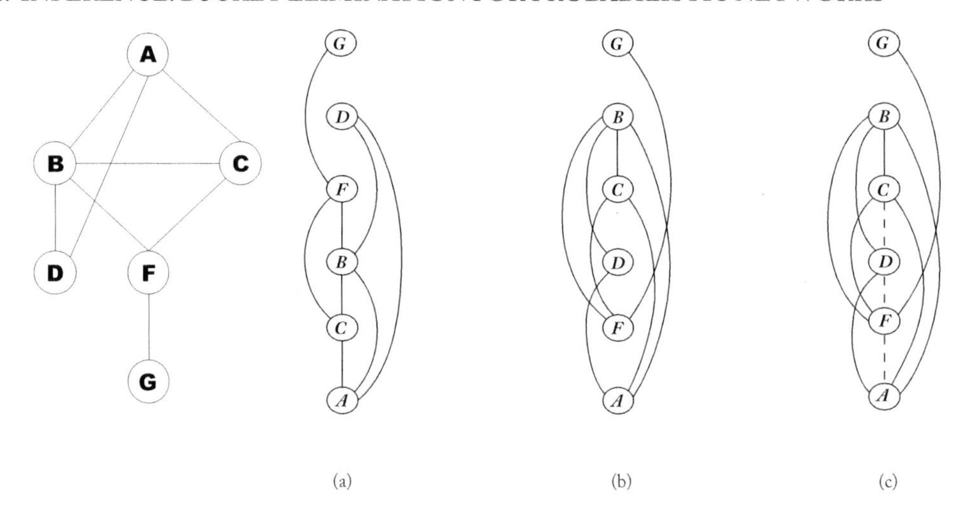

Figure 4.8: Two orderings, d_1 (a) and d_2 (b), of our example moral graph. In (c) the induced graph along ordering d_2.

$d_2 = A, F, D, C, B, G$, the ordered moral graph is depicted in Figure 4.8b and the induced graph is given in Figure 4.8c. In this ordering, the induced width is not the same as the width. For example, the width of C is initially 2, but its induced width is 3. The maximum induced width over all the variables in this ordering is 4 which is the induced-width of this ordering.

Theorem 4.6 Complexity of BE-bel. *Given a Byaesian network whose moral graph is G, let $w^*(d)$ be its induced width of G along ordering d, k the maximum domain size, and r be the number of input $CPTs$. The time complexity of BE-bel is $O(r \cdot k^{w^*(d)+1})$ and its space complexity is $O(n \cdot k^{w^*(d)})$ (see Appendix for a proof).*

4.1.3 THE IMPACT OF OBSERVATIONS

In this section we will see that observations, which are variable assignments, can have two opposing effects on the complexity of bucket-elimination BE-bel. One effect is of universal simplification and applies to any graphical model, while the other introduces complexity but is specific to likelihood queries over Bayesian networks.

Evidence removes connectivity

The presence of observed variables, which we call evidence in the Bayesian network context, is inherent to queries over probabilistic networks. From a computational perspective evidence is just an assignments of values to a subset of the variables. It turns out that the presence of such partial assignments can significantly simplify inference algorithms such as bucket elimination. In fact, we will see that this property of variable instantiations, or conditioning, as it is sometime called, is the basis for algorithms that combine search and variable-elimination, to be discussed in Chapter 7.

Take our belief network example with ordering d_1 and suppose we wish to compute the belief in A, having observed $B = b_0$. When the algorithm arrives at $bucket_B$, the bucket contains the three functions $P(b|a)$, $\lambda_D(b,a)$, and $\lambda_F(b,c)$, as well as the observation $B = b_0$ (see Figure 4.3 and add $B = b_0$ to $bucket_B$). Note that b_0 represent a specific value in the domain of B while b stands for an arbitrary value in its domain.

The processing rule dictates computing $\lambda_B(a,c) = P(b_0|a)\lambda_D(b_0,a)\lambda_F(b_0,c)$. Namely, generating and recording a two-dimensioned function. It would be more effective, however, to apply the assignment b_0 to each function in the bucket *separately* and then put the individual resulting functions into lower buckets. In other words, we can generate $\lambda_1(a) = P(b_0|a)$ and $\lambda_2(a) = \lambda_D(b_0,a)$, each of which has a single variable in its scope which will be placed in bucket A, and $\lambda_F(b_0,c)$, which will be placed in bucket C. By doing so, we avoid increasing the dimensionality of the recorded functions. In order to exploit this we introduce a special rule for processing buckets with observations (see step 3 in the algorithm): the observed value is assigned to each function in a bucket, and each function generated by this assignment is moved to the appropriate lower bucket.

Considering now ordering d_2, $bucket_B$ contains $P(b|a)$, $P(d|b,a)$, $P(f|c,b)$, and $B = b_0$ (see Figure 4.4a). The special rule for processing buckets holding observations will place the function $P(b_0|a)$ in $bucket_A$, $P(d|b_0,a)$ in $bucket_D$, and $P(f|c,b_0)$ in $bucket_F$. In subsequent processing only one-dimensional functions will be recorded. We see that in this case too the presence of observations reduces complexity: buckets of observed variables are processed in linear time and their recorded functions do not create functions on new subsets of variables.

Alternatively, we could just preprocess all the functions in which B appears and assign each the value b_0. This will reduce those functions scope and remove variable B altogether. We can then apply BE to the resulting pre-processed problem. Both methods will lead to an identical performance, but using an explicit rule for observations during *BE* allows for a more general and dynamic treatment. It can later be generalized by replacing observations by more general constraints (see Section 4.5).

In order to see the implication of the observation rule computationally, we can modify the way we manipulate the ordered moral graph and will not add arcs among parents of observed variables when computing the induced graph. This will permit a tighter bound on complexity. To capture this refinement we use the notion of *conditional induced graph*.

Definition 4.7 Conditional induced-width. Given a graph G, the conditional induced graph relative to ordering d and evidence variables $\mathbf{E}$, denoted $w^*_{\mathbf{E}}(d)$, is generated, processing the ordered graph from last to first, by connecting the earlier neighbors of unobserved nodes only.

The *conditional induced width* is the width of the conditional induced graph, disregarding observed nodes.

For example, in Figure 4.9a-b we show the ordered moral graph and induced ordered moral graph of the graph in Figures 4.1a–b. In Figure 4.9c the arcs connected to the observed node B are marked by broken lines and are disregarded, resulting in the conditional induced-graph. Modifying the complexity in Theorem 4.6, we get the following.

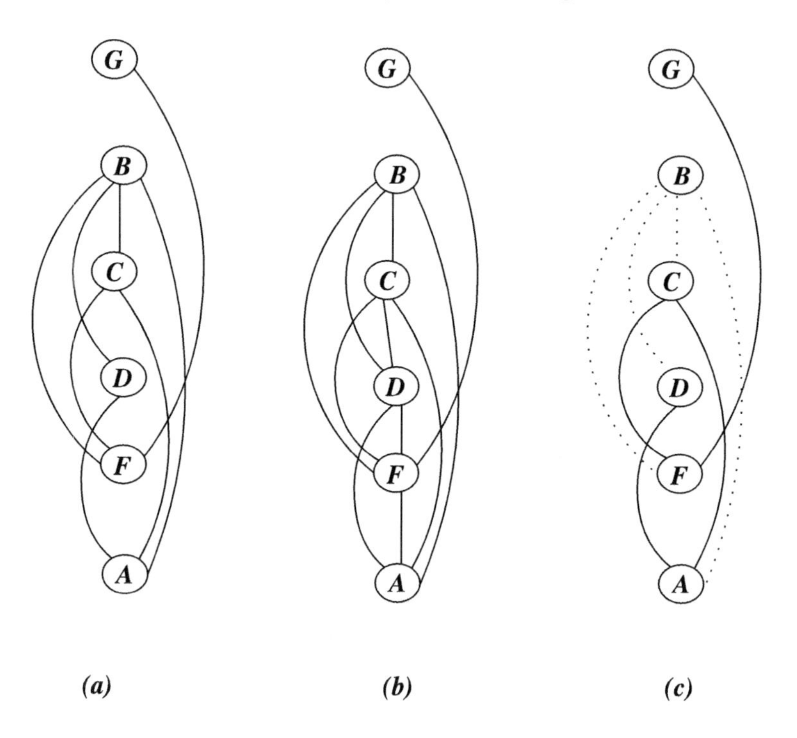

Figure 4.9: Adjusted induced graph relative to observing B.

Theorem 4.8 *Given a Bayesian network having n variables, algorithm BE-bel when using ordering d and evidence on variables* $\mathbf{E} = \mathbf{e}$, *is time and space exponential in the conditional induced width* $w^*_{\mathbf{E}}(d)$ *of the network's ordered moral graph. Specifically, its time complexity is* $O(r \cdot k^{w^*_{\mathbf{E}}(d)+1})$ *and its space complexity is* $O(n \cdot k^{w^*_{\mathbf{E}}(d)})$. □

It is easy to see that the conditional induced width is the induced width obtained by removing the evidence variables altogether (which correspond to the alternative preprocessing mentioned above).

Evidence creating connectivity: relevant subnetworks

We saw that observation can simplify computation. But, in Bayesian networks observation can also complicate inference. For example, when there is no evidence, computing the belief of a variable depends only on its non-descendant portion. This is because Bayesian networks functions are local probability distributions, and such functions, by definition, sum to 1. If we identify the portion of the network that is irrelevant, we can skip processing some of the buckets. For example, if we use a *topological ordering* from root to leaves along the directed acyclic graph (where parents precede their child nodes) and assuming that the queried variable is the first in the ordering, we can identify skippable buckets dynamically during processing.

Proposition 4.9 *Given a Bayesian network and a topological ordering $X_1, ..., X_n$, that begins a query variable X_1, algorithm BE-bel, computing $P(X_1|\mathbf{e})$, can skip a bucket if during processing the bucket contains no evidence variables and no newly computed messages.*

Proof: If topological ordering is used, each bucket of a variable X contains initially at most one function, $P(X|pa(X))$. Clearly, if there is neither evidence nor new functions in the bucket the summation operation $\sum_x P(x|pa(X))$ will yield the constant 1. □

Example 4.10 Consider again the belief network whose acyclic graph is given in Figure 4.1a and the ordering $d_1 = A, C, B, F, D, G$. Assume we want to update the belief in variable A given evidence on F. Obviously the buckets of G and D can be skipped and processing should start with $bucket_F$. Once $bucket_F$ is processed, the remaining buckets in the ordered processing are not skippable.

Alternatively, we can prune the non-relevant portion of the Bayesian network in advance, before committing to any processing ordering. The relevant subnetwork is called *ancestral subnetwork* and is defined recursively as follows.

Definition 4.11 Ancestral graph. Given a Bayesian network's directed graph $G = (X, E)$, and a query involving variables $\mathbf{S}$ (including the evidence variables), the ancestral graph of G, G_{anc}, relative to $\mathbf{S} \subseteq \mathbf{X}$, includes all variables in $\mathbf{S}$ and if a node is in G_{anc}, its parents are also in G_{anc}.

Example 4.12 Continuing with the example from Figure 4.1a, and assuming we want to assess the belief in A given evidence on F, the relevant ordered moral graph in Figures 4.1b should be modified by deleting nodes D and G. The resulting graph has nodes A, B, C, and F only.

Theorem 4.13 *Given a Bayesian $\mathcal{B} = \langle \mathbf{X}, \mathbf{D}, \mathbf{P}_G, \prod \rangle$ and a query $P(\mathbf{Y}|\mathbf{e})$, when $\mathbf{Y} \subseteq \mathbf{X}$ and $\mathbf{E} \subseteq \mathbf{X}$ is the evidence variable set, we can compute $P(\mathbf{Y}|\mathbf{e})$ by considering only the ancestral Bayesian network defined by $G_{\mathbf{Y} \cup \mathbf{E}}$. (Exercise: Prove the theorem.)*

4.2 BUCKET ELIMINATION FOR OPTIMIZATION TASKS

Belief-updating answers the question: "What is the likelihood of a variable given the observed data?" Answering that question, however, is often not enough; we want to be able to find the most likely explanation for the data we encounter. This is an optimization problem, and while we pose the problem here on a probabilistic network, it is a problem that is representative of optimization tasks on many types of graphical model. The query is called *mpe*.

The most probable explanation (mpe) task appears in numerous applications. Examples range from diagnosis and design of probabilistic codes to haplotype recognition in the context of family trees, and medical and circuit diagnosis. For example, given data on clinical findings, it may suggest the most likely disease a patient is suffering from. In decoding, the task is to identify the most likely input message which was transmitted over a noisy channel, given the observed output. Although the relevant task here is finding the most likely assignment over a *subset* of hypothesis variables which would correspond to a *marginal map* query, the mpe is close enough and is often used in applications (see [Darwiche, 2009] for more examples). Finally, the queries of mpe/map (see Chapter 2) drive most of the learning algorithms for graphical model [Koller and Friedman, 2009]. Our focus here is on algorithms for answering such queries on a given graphical model.

4.2.1 A BUCKET ELIMINATION ALGORITHM FOR MPE

Given a Bayesian network $\mathcal{B} = \langle \mathbf{X}, \mathbf{D}, \mathbf{P}_G, \prod \rangle$, the mpe task seeks an assignment to all the variables that has the maximal probability given the evidence. Namely, the task is to find a full instantiation $\mathbf{x}^0$ such that $P(\mathbf{x}^0) = \max_{\mathbf{x}} P(\mathbf{x}, \mathbf{e})$, where denoting $\mathbf{x} = (x_1, ..., x_n)$, $P(\mathbf{x}, \mathbf{e}) = \prod_i P(x_i, \mathbf{e}|\mathbf{x}_{pa_i})$. (Remember the x_{pa_i} is the assignments to x restricted to the variables in the parent set of X_i.) Given a variable ordering $d = X_1, ..., X_n$, we can accomplish this task by performing maximization operation, variable by variable, along the ordering from last to first (i.e., right to left), migrating to the left all CPTs that do not mention the maximizing variable. We will derive this algorithm in a similar way to that in which we derived BE-bel. Using the notation defined earlier for operations on functions, our goal is to find M, s.t.

$$M = \max_{\mathbf{x}} P(\mathbf{x}_{(1..n)}, \mathbf{e}) = \max_{\mathbf{x}_{(1..n-1)}} \max_{x_n} \prod_i P(x_i, \mathbf{e}|\mathbf{x}_{pa_i})$$

$$= \max_{\mathbf{x}_{(1..n-1)}} \prod_{X_i \in X - M_n} P(x_i, \mathbf{e}|\mathbf{x}_{pa_i}) \cdot \max_{x_n} P(x_n, \mathbf{e}|\mathbf{x}_{pa_n}) \prod_{X_i \in M_n} P(x_i, \mathbf{e}|\mathbf{x}_{pa_i})$$

$$= \max_{\mathbf{x}_{(1..n-1)}} \prod_{X_i \in X - M_n} P(x_i, \mathbf{e}|\mathbf{x}_{pa_i}) \cdot h_n(\mathbf{x}_{S_n})$$

where

$$h_n(\mathbf{x}_{S_n}) = \max_{x_n} P(x_n, \mathbf{e}|\mathbf{x}_{pa_n}) \prod_{X_i \in M_n} P(x_i, \mathbf{e}|\mathbf{x}_{pa_i})$$

and S_n is the scope of the generated function h_n, and M_n is the Markov blanket of X_n. Clearly, the algebraic manipulation of the above expressions is the same as the algebraic manipulation for belief updating where summation is replaced by maximization. Consequently, the bucket elimination procedure *BE-mpe* is identical to BE-bel except for this change in the margnalization operator.

Given ordering $d = (X_1, ..., X_n)$, the conditional probability tables are partitioned as before. To process each bucket, we take the product of all the functions that reside in the bucket and then eliminate the bucket's variable by maximization. We distinguish again between the original function in the bucket whose product is denoted ψ_p and the messages, which in this case will be denoted by h. The generated function in the bucket of X_p is $h_p : S_p \rightarrow R$, $h_p = \max_{X_p} \psi_p \cdot \prod_{i=1}^{j} h_i$, where $S_p = scope(\psi_p) \cup \cup_i scope(h_i) - \{X_p\}$, is the order restricted Markov blanket and it is placed in the bucket of its largest-index variable in S_p. If the function is a constant, we can place it directly in the first bucket; constant functions are not necessary to determine the exact mpe value.

Bucket processing continues from the last to the first variable. Once all buckets are processed, the mpe value can be extracted as the maximizing product of functions in the first bucket. At this point we know the mpe value but we have not generated an optimizing tuple (also called configuration). The algorithm initiates a *forward phase* to compute an mpe tuple by assigning values to the variables along the ordering from X_1 to X_n, consulting the information recorded in each bucket. Specifically, the value x_i of X_i is selected to maximize the product in $bucket_i$ given the partial assignment $\mathbf{x}_{(1..(i-1))} = (x_1, ..., x_{i-1})$. The algorithm is presented in Figure 4.10. Observed variables are handled as in BE-bel.

Example 4.14 Consider again the belief network in Figure 4.1a. Given the ordering $d = A, C, B, F, D, G$ and the evidence $G = 1$, we process variables from last to first once partitioning the conditional probability functions into buckets, as was shown in Figure 4.2 To process G, assign $G = 1$, get $h_G(f) = P(G = 1|f)$ and place the result in $bucket_F$. We next process $bucket_D$ by computing $h_D(b, a) = \max_d P(d|b, a)$ and put the result in $bucket_B$. Bucket F, which is next to be processed, now contains two functions: $P(f|b, c)$ and $h_G(f)$. We Compute $h_F(b, c) = \max_f p(f|b, c) \cdot h_G(f)$, and place the resulting function in $bucket_B$. To process $bucket_B$, we record the function $h_B(a, c) = \max_b P(b|a) \cdot h_D(b, a) \cdot h_F(b, c)$ and place it in $bucket_C$. To process C (to eliminate C), we compute $h_C(a) = \max_c P(c|a) \cdot h_B(a, c)$ and place it in $bucket_A$. Finally, the mpe value given in $bucket_A$, $M = \max_a P(a) \cdot h_C(a)$, is determined. Next, the mpe configuration is generated by going forward through the buckets. First, the value a^0 satisfying $a^0 = argmax_a P(a) h_C(a)$ is selected. Subsequently, the value of C, $c^0 = argmax_c P(c|a^0) h_B(a^0, c)$ is determined. Next, $b^0 = argmax_b P(b|a^0) h_D(b, a^0) h_F(b, c^0)$ is selected, and so on. The schematic computation is the same as in Figure 4.3 where λ is simply replaced by h.

Algorithm BE-mpe
Input: A belief network $\mathcal{B} = \langle \mathbf{X}, \mathbf{D}, \mathbf{P}_G, \prod \rangle$, where $\mathcal{P} = \{P_1, ..., P_n\}$; an ordering of the variables, $d = X_1, ..., X_n$; observations $\mathbf{e}$.
Output: The most probable configuration given the evidence.
1. **Initialize:** Generate an ordered partition of the conditional probability function, $bucket_1, \ldots,$ $bucket_n$, where $bucket_i$ contains all functions whose highest variable is X_i. Put each observed variable in its bucket. Let ψ_i be the product of input function in a bucket and let h_i be the messages in the bucket.

2. **Backward:** For $p \leftarrow n$ downto 1, do
for all the functions $h_1, h_2, ..., h_j$ in $bucket_p$, do

- **If** (observed variable) $bucket_p$ contains $X_p = x_p$, assign $X_p = x_p$ to each function and put each in appropriate bucket.
- **else**, Generate functions $h_p \Leftarrow \max_{X_p} \psi_p \cdot \Pi_{i=1}^{j} h_i$
Add h_p to the bucket of the largest-index variable in $scope(h_p)$.

3. **Forward:**

- Generate the mpe cost by maximizing over X_1, the product in $bucket_1$. Namely $mpe = max_{X_1} \psi_1 \prod_j h_{1_j}$.
- (generate an mpe tuple)
For $i = 1$ to n along d do: Given $\mathbf{x}_{(1..(i-1))} = (x_1, ..., x_{i-1})$ Choose $x_i^o = argmax_{X_i} \psi_i \cdot \Pi_{\{h_j \in \, bucket_i\}} h_j(\mathbf{x}_{(1..(i-1))})$.

4. **Output:** mpe and configuration $\mathbf{x}^o$.

Figure 4.10: Algorithm *BE-mpe.*

The backward process can be viewed as a compilation phase in which we compile information regarding the most probable extension (cost to go) of partial tuples to variables higher in the ordering.

Complexity. As in the case of belief updating, the complexity of BE-mpe is bounded exponentially by the arity of the recorded functions, and those functions depend on the induced width and the evidence

Theorem 4.15 Soundness and Complexity. *Algorithm BE-mpe is complete for the mpe task. Its time and space complexity are $O(r \cdot k^{w^*_{\mathbf{E}}(d)+1})$ and $O(n \cdot k^{w^*_{\mathbf{E}}(d)})$, respectively, where n is the number*

of variables, k bound the domain size and $w^*_{\mathbf{E}}(d)$ is the induced width of the ordered moral graph along d, conditioned on the evidence **E**. □

4.2.2 A BUCKET ELIMINATION ALGORITHM FOR MAP

The maximum a'posteriori hypothesis *map*[1] task is a generalization of both mpe and belief updating. It asks for the maximal probability associated with a *subset of hypothesis variables* and is widely applicable especially for diagnosis tasks. Belief updating is the special case where the hypothesis variables are just single variables. The mpe query is the special case when the hypothesis variables are all the unobserved variables. We will see that since it is a mixture of the previous two tasks, in its bucket-elimination algorithm some of the variables are eliminated by summation while others by maximization.

Given a Bayesian network, a subset of hypothesized variables $A = \{A_1, ..., A_k\}$, and some evidence e, the problem is to find an assignment (i.e., configuration) to A having maximum probability given the evidence compared with all other assignments to A. Namely, the task is to find $a^o = \text{argmax}_{a_1,...,a_k} P(a_1, ..., a_k, \mathbf{e})$ (see also Definition 2.22). So, we wish to compute $\max_{\mathbf{a}_{1..k}} P(a_1, ..., a_k, e) = \max_{\mathbf{a}_{1..k}} \sum_{\mathbf{x}_{(k+1..n)}} \prod_{i=1}^{n} P(x_i, \mathbf{e}|\mathbf{x}_{pa_i})$ where $\mathbf{x} = (a_1, ..., a_k, x_{k+1}, ..., x_n)$. Algorithm *BE-map* in Figure 4.11 considers only orderings in which the hypothesized variables start the ordering because summation should be applied first to the subset of variables which are in $\mathbf{X} - \mathbf{A}$, and subsequently maximization is applied to the variables in $\mathbf{A}$. Since summation and maximization cannot be permuted we have to be restricted in the orderings. Like BE-mpe, the algorithm has a backward phase and a forward phase, but the forward phase is restricted to the hypothesized variables only. Because only restricted orderings are allowed, the algorithm may be forced to have far higher induced-width than would otherwise be allowed.

Theorem 4.16 *Algorithm BE-map is complete for the map task for orderings started by the hypothesis variables. Its time and space complexity are $O(r \cdot k^{w^*_{\mathbf{E}}(d)+1})$ and $O(n \cdot k^{w^*_{\mathbf{E}}(d)})$, respectively, where n is the number of variables in graph, k bounds the domain size and $w^*_{\mathbf{E}}(d)$ is the conditioned induced width of its moral graph along d, relative to evidence variables* **E**. *(Prove as an exercise.)* □

4.3 BUCKET ELIMINATION FOR MARKOV NETWORKS

Recalling Definition 2.23 of a Markov network which is presented here for convenience.

Definition 4.17 Markov networks. A Markov network is a graphical model $\mathcal{M} = \langle \mathbf{X}, \mathbf{D}, \mathbf{H}, \prod \rangle$ where $\mathbf{H} = \{\psi_1, \ldots, \psi_m\}$ is a set of potential functions where each potential ψ_i

[1] Sometimes map is meant to refer to the mpe, and the map task is called marginal map.

Algorithm BE-map

Input: A Bayesian network $\mathcal{B} = \langle \mathbf{X}, \mathbf{D}, \mathbf{P}_G, \prod \rangle$, $P = \{P_1, ..., P_n\}$; a subset of hypothesis variables $A = \{A_1, ..., A_k\}$; an ordering of the variables, d, in which the A's are first in the ordering; observations e. ψ_i is the product of input function in the bucket of X_i.

Output: A most probable assignment $A = a$.

1. **Initialize:** Generate an ordered partition of the conditional probability functions, $bucket_1$, ..., $bucket_n$, where $bucket_i$ contains all functions whose highest variable is X_i.
2. **Backwards** For $p \leftarrow n$ downto 1, do

for all the message functions $\beta_1, \beta_2, ..., \beta_j$ in $bucket_p$ and for ψ_p do

- **If** (observed variable) $bucket_p$ contains the observation $X_p = x_p$, assign $X_p = x_p$ to each β_i and ψ_p and put each in appropriate bucket.
- **else,** If X_p is not in A, then $\beta_p \Leftarrow \sum_{X_p} \psi_p \cdot \Pi_{i=1}^{j} \beta_i$;
  else, $(X_p \in A)$, $\beta_p \Leftarrow \max_{X_p} \psi_p \cdot \prod_{i=1}^{j} \beta_i$
  Place β_p in the bucket of the largest-index variable in $scope(\beta_p)$.

3. **Forward:** Assign values, in the ordering $d = A_1, ..., A_k$, using the information recorded in each bucket in a similar way to the forward pass in BE-mpe.
4. **Output:** Map and the corresponding configuration over A.

Figure 4.11: Algorithm *BE-map*.

is a non-negative real-valued function defined over a scope of variables $\mathbf{S}_i$. The Markov network represents a global joint distribution over the variables $\mathbf{X}$ given by:

$$P(\mathbf{x}) = \frac{1}{Z} \prod_{i=1}^{m} \psi_i(\mathbf{x}) \quad , \quad Z = \sum_{\mathbf{x}} \prod_{i=1}^{m} \psi_i(\mathbf{x})$$

where the normalizing constant Z is referred to as the partition function.

It is easy to see that the bucket-elimination algorithms we presented for Bayesian networks are immediately applicable to all the main queries over Markov networks. All we need to do is replace the input conditional probabilities through which a Bayesian network is specified by the collection of local potential functions or factors denoted by $\psi(.)$. The query of computing posterior marginals is accomplished by BE-bel, computing mpe and map are accomplished by BE-mpe and BE-map, respectively. Since the partition function is identical, mathematically to the expression of probability of evidence, the task of computing Z is identical algorithmically to the task of computing the probability of the evidence.

4.4 BUCKET ELIMINATION FOR COST NETWORKS AND DYNAMIC PROGRAMMING

As we mentioned at the outset, bucket-elimination algorithms are variations of a very well known class of optimization algorithms known as *Dynamic Programming* [Bellman, 1957; Bertele and Brioschi, 1972]. Here we make the connection explicit, observing that BE-mpe is a dynamic programming scheme with some simple transformations.

That BE-mpe is dynamic programming becomes apparent once we transform the mpe's cost function, which has a product combination operator, into the traditional additive combination operator using the log function. For example,

$$P(a,b,c,d,f,g) = P(a)P(b|a)P(c|a)P(f|b,c)P(d|a,b)P(g|f)$$

becomes

$$C(a,b,c,d,e) = -logP = C(a) + C(b,a) + C(c,a) + C(f,b,c) + C(d,a,b) + C(g,f)$$

where each $C_i = -logP_i$.

The general dynamic programming algorithm is defined over *cost networks* (see Section 2.4). As we showed a *cost network* is a tuple $C = \langle \mathbf{X}, \mathbf{D}, \mathbf{F}, \sum \rangle$, where $\mathbf{X} = \{X_1, ..., X_n\}$ are variables over domains $\mathbf{D} = \{D_1, ..., D_n\}$, F is a set of real-valued cost functions $C_1, ..., C_l$, defined over scopes $S_1, ..., S_l$. The task is to find an assignment or a configuration to all the variables that minimizes the global function $\sum_i C_i$.

A straightforward elimination process similar to that of BE-mpe (where the product is replaced by summation and maximization by minimization) yields the non-serial dynamic programming algorithm in [Bertele and Brioschi, 1972]. The algorithm, called here *BE-opt*, is given in Figure 4.12. Evidence is not assumed to be part of the input, so this part of the algorithm is omitted.

A schematic execution of our example along ordering $d = G, A, F, D, C, B$ is depicted in Figure 4.13. It is identical to what we saw for BE-mpe, except that the generated functions are computed by min-sum, instead of max-product. Not surprisingly, we can show the following.

Theorem 4.18 *Given a cost network $C = \langle \mathbf{X}, \mathbf{D}, \mathbf{F}, \sum \rangle$, BE-opt is complete for finding an optimal cost solution. Its time and space complexity are $O(r \cdot k^{w^*(d)+1})$ and $O(n \cdot k^{w^*(d)})$, respectively, where n is the number of variables in graph, k bounds the domain size, and $w^*(d)$ is the induced width of its primal graph along d.* □

Consulting again the various classes of cost networks elaborated up on in Section 2.4, algorithm, BE-opt is applicable to all including weighted-csps, max-csps and max-sat.

Algorithm BE-opt
Input: A cost network $C = \langle \mathbf{X}, \mathbf{D}, \mathbf{F}, \sum \rangle$, $F = \{C_1, ..., C_l\}$; ordering d.
Output: A minimal cost assignment.
1. **Initialize:** Partition the cost components into buckets. Define ψ_i as the sum of the input cost functions in bucket X_i.
2. **Process buckets** from $p \leftarrow n$ downto 1
For ψ_p and the cost messages $h_1, h_2, ..., h_j$ in $bucket_p$, do:

- (sum and minimize):
  $h_p \Leftarrow min_{X_p}(\psi_p + \sum_{i=1}^{j} h_i)$. Place h_p into the largest index variable in its scope.

3. **Forward:** Assign minimizing values in ordering d, consulting functions in each bucket (as in BE-mpe).
4. **Return:** Optimal cost and an optimizing assignment.

Figure 4.12: Dynamic programming as BE-opt.

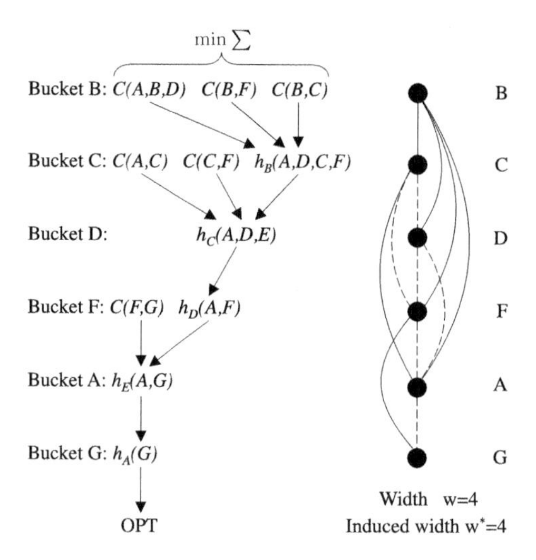

Figure 4.13: Schematic execution of BE-opt.

4.5 BUCKET ELIMINATION FOR MIXED NETWORKS

The last class of graphical models we will address is mixed network defined in Section 2.6. To refresh, these models allow the explicit representation of both probabilistic information and constraints. The mixed network is defined by a pair of a Bayesian network and a constraint network. This pair expresses a probability distribution over all the variables which is conditioned on the requirement that all the assignments having non-zero probability satisfy all the constraints.

We will focus only on the task that is unique to this graphical model, the *constraint probability evaluation* (CPE), which can also stand for *CNF probability evaluation*. Given a mixed network $\mathcal{M}_{(\mathcal{B},\varphi)}$, where φ is a CNF formula defined on perhaps a subset of propositional variables and $\mathcal{B}$ is a Bayesian network, the CPE task is to compute $P(\mathbf{x} \in Mod(\varphi))$, where the models of φ are denoted by $Mod(\varphi)$ and are the assignments to all the variables satisfying the formula φ. We denote by $P_{\mathcal{B}}(\varphi)$ the probability that $P_{\mathcal{B}}(\mathbf{x} \in Mod(\varphi))$. We denote by $\mathbf{X}_\varphi$ the set of variables in the scope of φ. By definition,

$$P_{\mathcal{B}}(\varphi) = \sum_{\mathbf{x}_\varphi \in Mod(\varphi)} P(\mathbf{x}_\varphi)$$

Using the belief network product form we get:

$$P_{\mathcal{B}}(\varphi) = \sum_{\{\mathbf{x} | \mathbf{x}_\varphi \in Mod(\varphi)\}} \prod_{i=1}^{n} P(x_i | \mathbf{x}_{pa_i}).$$

We separate the summation over X_n and the rest of the variables $\mathbf{X} - \{X_n\}$ as usual and denote by φ_{+X_n} the set of all clauses defined on X_n (there may be none if it is not a proposition in φ) and by φ_{-X_n} all the rest of the clauses which are not defined on X_n. We get (denoting $P(X_i|\mathbf{x}_{pa_i})$ by P_i):

$$P_{\mathcal{B}}(\varphi) = \sum_{\{\mathbf{x}_{(1..n-1)} | \mathbf{x}_{\varphi - X_n} \in Mod(\varphi_{-X_n})\}} \sum_{\{x_n | \mathbf{x}_{\varphi + X_n} \in Mod(\varphi_{+X_n})\}} \prod_{i=1}^{n} P(x_i | \mathbf{x}_{pa_i}).$$

Let t_n be the set of indices of functions in the product that *do not* mention X_n, namely, are not in φ_{+X_n} and by $l_n = \{1, \ldots, n\} \setminus t_n$ we get:

$$P_{\mathcal{B}}(\varphi) = \sum_{\{\mathbf{x}_{(1..n-1)} | \mathbf{x}_{\varphi - X_n} \in Mod(\varphi_{-X_n})\}} \prod_{j \in t_n} P_j \cdot \sum_{\{x_n | \mathbf{x}_{\varphi + X_n} \in Mod(\varphi_{+X_n})\}} \prod_{j \in l_n} P_j.$$

Therefore:

$$P_{\mathcal{B}}(\varphi) = \sum_{\{\mathbf{x}_{(1..n-1)} | \mathbf{x}_{\varphi - X_n} \in Mod(\varphi_{-X_n})\}} (\prod_{j \in t_n} P_j) \cdot \lambda_{X_n},$$

where λ_{X_n} is defined over $U_n = scope(\varphi_{+X_n})$, by

$$\lambda_{X_n} = \sum_{\{x_n | \mathbf{x}_{U_n} \in Mod(\varphi_{+X_n})\}} \prod_{j \in l_n} P_j. \tag{4.8}$$

The case of observed variables. When X_n is observed, that is constrained by a literal, the summation operation reduces to assigning the observed value to each of its CPTs *and* to each of the relevant clauses. In this case, Eq. (4.8) becomes (assume $X_n = x_n$ and $P_{(=x_n)}$ is the function instantiated by assigning x_n to X_n):

$$\lambda_{X_n} = \prod_{j \in l_n} P_{j(=x_n)}, \quad if\ \mathbf{x}_{U_n} \in Mod(\varphi_{+X_n}) \wedge (X_n = x_n)). \tag{4.9}$$

Algorithm 1: BE-CPE

Input: A belief network $\mathcal{M} = (\mathcal{B}, \simeq)$, $\mathcal{B} = \langle \mathbf{X}, \mathbf{D}, \mathbf{P}_G, \prod \rangle$, where $\mathcal{B} = \{P_1, ..., P_n\}$; a CNF formula on k propositions $\varphi = \{\alpha_1, ...\alpha_m\}$ defined over k propositions; an ordering of the variables, $d = \{X_1, \ldots, X_n\}$.

Output: The belief $P(\varphi)$.

1 Place buckets with unit clauses last in the ordering (to be processed first).

`// Initialize`

Partition $\mathcal{B}$ and φ into $bucket_1, \ldots, bucket_n$, where $bucket_i$ contains all the CPTs and clauses whose highest variable is X_i.

Put each observed variable into its appropriate bucket. (We denote probabilistic functions by λs and clauses by αs).

2 **for** $p \leftarrow n$ **downto** 1 **do** `// Backward`

  Let $\lambda_1, \ldots, \lambda_j$ be the functions and $\alpha_1, \ldots, \alpha_r$ be the clauses in $bucket_p$

  `Process-bucket`$_p(\sum, (\lambda_1, \ldots, \lambda_j),(\alpha_1, \ldots, \alpha_r))$

3 **return** $P(\varphi)$ as the result of processing $bucket_1$.

Otherwise, $\lambda_{X_n} = 0$. Since a tuple $\mathbf{x}_{U_n}$ satisfies $\varphi_{+X_n} \wedge (X_n = x_n)$ only if $\mathbf{x}_{U_n - X_n}$ satisfies the resolvent clause $\gamma_n = resolve(\varphi_{+X_n}, (X_n = x_n))$, we get:

$$\lambda_{X_n} = \prod_{j \in l_n} P_{j(=x_n)}, \quad if \mathbf{x}_{(U_n - X_n)} \in Mod(\gamma_n). \tag{4.10}$$

We can, therefore, extend the case of observed variable in a natural way: CPTs are assigned the observed value as usual while clauses are individually resolved with the unit clause $(X_n = x_n)$, and both are moved to appropriate lower buckets. This yields the following.

To Initialise, place all CPTs and clauses mentioning X_n in its bucket and then compute the function in Eq. (4.8). The computation of the rest of the expression proceeds with X_{n-1} in the same manner. This yields algorithm BE-CPE described in Figure 1 and Procedure `Process-bucket`$_p$. The elimination operation is summation for the current query. Thus, for every ordering of the propositions, once all the CPTs and clauses are partitioned, we process the buckets from last to first, in each applying the following operation. Let $\lambda_1, ...\lambda_t$ be the probabilistic functions in $bucket_P$ and $\varphi = \{\alpha_1, ...\alpha_r\}$ be the clauses. The algorithm computes a new function λ_P over $S_p = scope(\lambda_1, ...\lambda_t) \cup scope(\alpha_1, ...\alpha_r) - \{X_p\}$ defined by:

$$\lambda_P = \sum_{\{x_p | \mathbf{x}_\varphi \in Mod(\alpha_1, \ldots, \alpha_r)\}} \prod_j \lambda_j .$$

Example 4.19 Consider the belief network in Figure 4.14, which is similar to the one in Figure 2.5, and the query $\varphi = (B \vee C) \wedge (G \vee D) \wedge (\neg D \vee \neg B)$. The initial partitioning into buckets

Procedure `Process-bucket`$_p$($\sum$, $(\lambda_1, \ldots, \lambda_j)$,$(\alpha_1, \ldots, \alpha_r)$).

if *bucket*$_p$ *contains evidence* $X_p = x_p$ **then**

1. Assign $X_p = x_p$ to each λ_i and put each resulting function in the bucket of its latest variable
2. Resolve each α_i with the unit clause, put non-tautology resolvents in the buckets of their latest variable and **move any bucket with unit clause to top of processing**

else

$\lambda_p \leftarrow \sum_{\{x_p | \mathbf{x}_{U_p} \in Mod(\alpha_1, \ldots, \alpha_r)\}} \prod_{i=1}^{j} \lambda_i$

Add λ_p to the bucket of the latest variable in S_p, where

$S_p = scope(\lambda_1, \ldots, \lambda_j, \alpha_1, \ldots, \alpha_r)$, $U_p = scope(\alpha_1, \ldots, \alpha_r)$.

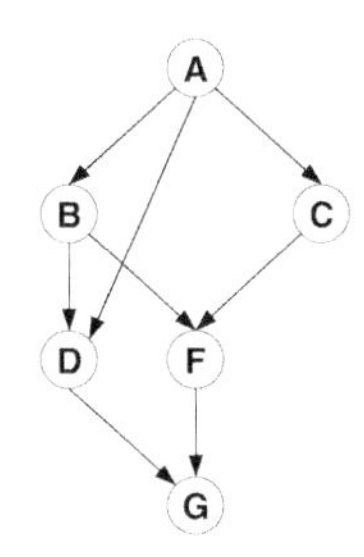

(a) Directed acyclic graph

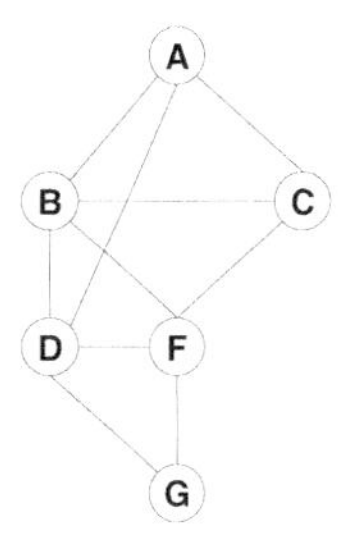

(b) Moral graph

Figure 4.14: Belief network.

along the ordering $d = A, C, B, D, F, G$, as well as the output buckets are given in Figure 4.15. We compute:

In Bucket G: $\quad \lambda_G(f, d) = \sum_{\{g | g \vee d = true\}} P(g|f)$

In $Bucket_F$: $\quad \lambda_F(b, c, d) = \sum_f P(f|b, c)\lambda_G(f, d)$

In $Bucket_D$: $\quad \lambda_D(a, b, c) = \sum_{\{d | \neg d \vee \neg b = true\}} P(d|a, b)\lambda_F(b, c, d)$

In $Bucket_B$: $\quad \lambda_B(a, c) = \sum_{\{b | b \vee c = true\}} P(b|a)\lambda_D(a, b, c)\lambda_F(b, c)$

In $Bucket_C$: $\quad \lambda_C(a) = \sum_c P(c|a)\lambda_B(a, c)$

In $Bucket_A$: $\quad \lambda_A = \sum_a P(a)\lambda_C(a)$

$P(\varphi) = \lambda_A$.

For example in $bucket_G$, $\lambda_G(f, d = 0) = P(g = 1|f)$, because if $D = 0$ g must get the value "1", while $\lambda_G(f, d = 1) = P(g = 0|f) + P(g = 1|f)$. In summary, we have the following.

Theorem 4.20 Correctness and completeness. *Algorithm BE-cpe is sound and complete for the CPE task.*

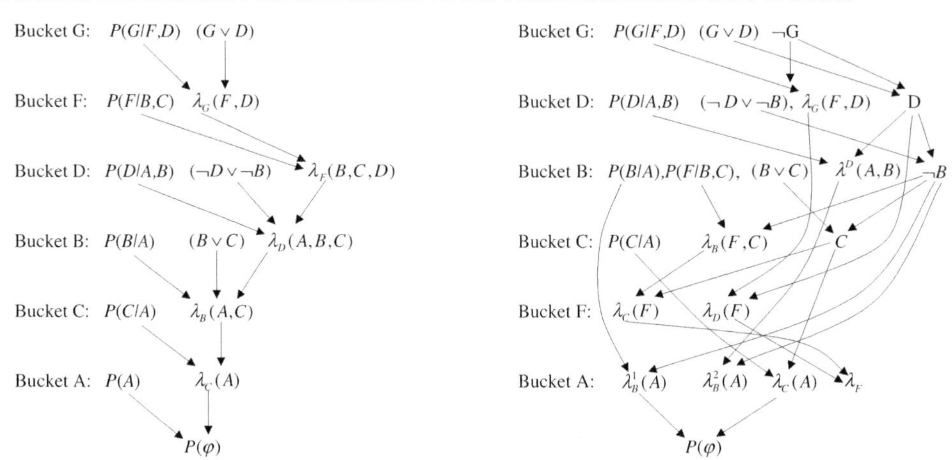

Figure 4.15: Execution of BE-CPE.

Figure 4.16: Execution of BE-CPE (evidence $\neg G$).

It is easy to see that the complexity of the algorithm depends on the mixed graph which is the union of the moral graph and the constraint graph, in the usual way.

Theorem 4.21 Complexity of BE-cpe. *Given a mixed network $M_{\mathcal{B},\varphi}$ having mixed graph is G, with $w^*(d)$ its induced width along ordering d, k the maximum domain size and r be the number of input functions. The time complexity of BE-cpe is $O(r \cdot k^{w^*(d)+1})$ and its space complexity is $O(n \cdot k^{w^*(d)})$. (Prove as an exercise.)*

Notice that algorithm BE-cpe also includes a unit resolution step whenever possible (see Procedure $\texttt{Process-bucket}_p$) and a dynamic reordering of the buckets that prefers processing buckets that include unit clauses. This may have a significant impact on efficiency because treating observations (namely unit clauses) in a special way can avoid creating new dependencies as we already observed.

Example 4.22 Let's now extend the example by adding $\neg G$ to the query. This will place $\neg G$ in the bucket of G. When processing bucket G, unit resolution creates the unit clause D, which is then placed in bucket D. Next, processing bucket F creates a probabilistic function on the two variables B and C. Processing bucket D that now contains a unit clause will assign the value $D = 1$ to the CPT in that bucket and apply unit resolution, generating the unit clause $\neg B$ that is placed in bucket B. Subsequently, in bucket B we can apply unit resolution again, generating C placed in bucket C, and so on. In other words, aside from bucket F, we were able to process all buckets as observed buckets, by propagating the observations (see Figure 4.16.) To incorporate dynamic variable ordering, after processing bucket G, we move bucket D to the top of the processing list (since it has a unit clause). Then, following its processing, we process bucket B and then bucket C, then F, and finally A.

Since unit resolution increases the number of buckets having unit clauses, and since those are processed in linear time, it can improve performance substantially. Such buckets can be identified a priori by applying unit resolution on the CNF formula or arc-consistency if we have a constraint expression. In fact, any level of resolution can be applied in each bucket. This can yield stronger CNF expressions in each bucket and may help improve the computation of the probabilistic functions (see [Dechter, 2003]).

4.6 THE GENERAL BUCKET ELIMINATION

We now summarize and generalize the bucket elimination algorithm using the two operators of *combination* and *marginalization*. As presented in Chapter 2, the general task can be defined over a graphical model $\mathcal{M} = \langle \mathbf{X}, \mathbf{D}, \mathbf{F}, \otimes \rangle$, where: $X = \{X_1, ..., X_n\}$ is a set of variables having domain of values $D = \{D_1, ..., D_n\}$ and $F = \{f_1, ..., f_k\}$ is a set of functions, where each f_i is defined over $S_i = scope(f_i)$. Given a function h and given $Y \subseteq scope(h)$, the (generalized) projection operator $\Downarrow_Y h$, $\Downarrow_Y h \in \{max_{S-Y} h, min_{S-Y} h, \pi_Y h, \sum_{S-Y} h\}$ and the (generalized) combination operator $\otimes_j f_j$ defined over $U = \cup_j scope(f_j)$, $\otimes_{j=1}^k f_j \in \{\Pi_{j=1}^k f_j, \sum_{j=1}^k f_j, \bowtie_j f_j\}$. All queries require computing $\Downarrow_Y \otimes_{i=1}^n f_i$. Such problems can be solved by a general bucket-elimination algorithm stated in Figure 4.17. For example, BE-bel is obtained when $\Downarrow_Y = \sum_{S-Y}$ and $\otimes_j = \Pi_j$, BE-mpe is obtained when $\Downarrow_Y = \max_{S-Y}$ and $\otimes_j = \Pi_j$, and adaptive consistency corresponds to $\Downarrow_Y = \pi_Y$ and $\otimes_j = \bowtie_j$. Similarly, Fourier elimination and directional resolution can be shown to be expressible in terms of such operators. For mixed networks the combination and marginalization are also well defined.

We will state briefly the properties of GBE.

Theorem 4.23 Correctness and complexity. *Algorithm GBE is sound and complete for its task. Its time and space complexities is exponential in the $w^*(d) + 1$ and $w^*(d)$, respectively, along the order of processing d.*

4.7 SUMMARY AND BIBLIOGRAPHICAL NOTES

In the last two chapters, we showed how the bucket-elimination framework can be used to unify variable-elimination algorithms for both deterministic and probabilistic graphical models for various tasks. The algorithms take advantage of the structure of the graph. Most bucket-elimination algorithms are time and space exponential in the induced width of the underlying dependency primal graph of the problem.

Chapter 4 is based on Dechter's Bucket-elimination algorithm that appeared in [Dechter, 1999]. Among the early variable elimination algorithms we find the peeling algorithm for genetic

Algorithm General bucket elimination (GBE)

Input: $\mathcal{M} = \langle \mathbf{X}, \mathbf{D}, \mathbf{F}, \otimes \rangle$. $F = \{f_1, ..., f_n\}$ an ordering of the variables, $d = X_1, ..., X_n$; $\mathbf{Y} \subseteq \mathbf{X}$.

Output: A new compiled set of functions from which the query $\Downarrow_Y \otimes_{i=1}^n f_i$ can be derived in linear time.

1. **Initialize:** Generate an ordered partition of the functions into $bucket_1, ..., bucket_n$, where $bucket_i$ contains all the functions whose highest variable in their scope is X_i. An input function in each bucket ψ_i, $\psi_i = \otimes_{i=1}^n f_i$.
2. **Backward:** For $p \leftarrow n$ downto 1, do
for all the functions $\psi_p, \lambda_1, \lambda_2, ..., \lambda_j$ in $bucket_p$, do

- **If** (observed variable) $X_p = x_p$ appears in $bucket_p$, assign $X_p = x_p$ in ψ_p and to each λ_i and put each resulting function in appropriate bucket.
- **else,** (combine and marginalize)
$\lambda_p \leftarrow \Downarrow_{S_p} \psi_p \otimes (\otimes_{i=1}^j \lambda_i)$ and add λ_p to the largest-index variable in $scope(\lambda_p)$.

3. **Return:** all the functions in each bucket.

Figure 4.17: Algorithm *General bucket elimination.*

trees [Cannings *et al.*, 1978], Zhang and Poole's VE1 algorithm [Zhang and Poole, 1996], and SPI algorithm by D'Ambrosio et al., [R.D. Shachter and Favero, 1990] which preceded both BE-bel and VE1 and provided the principle ideas in the context of belief updating. Decimation algorithms in statistical physics are also related and were applied to Boltzmann trees [Saul and Jordan, 1994].

In [R. Dechter and Pearl, 1990] the connection between optimization and constraint satisfaction and its relationship to dynamic programming is explicated. In the work of [Mitten, 1964; Shenoy, 1992] and later in [Bistarelli *et al.*, 1997] an axiomatic framework that characterize tasks that can be solved polynomially over hyper-trees, is introduced.

4.8 APPENDIX: PROOFS

Proof of Theorem 4.6

During BE-bel, each bucket creates a λ function which can be viewed as a message that it sends to a *parent* bucket, down the ordering (recall that we process the variables from last to first). Since to compute this function over w^* variables the algorithm needs to consider all the tuples defined on all the variables in the bucket, whose number is bounded by $w^* + 1$, the time to compute the function is bounded by k^{w^*+1}, and its size is bounded by k^{w^*}. For each of these k^{w^*+1} tuple we need to compute its value by considering information from each of the functions in the buckets. If r_i is the number of the bucket's original messages and deg_i is the number of messages it receives from its children, then the computation of the bucket's function is $O((r_i + deg_i + 1)k^{w^*+1})$. Therefore, summing over all the buckets, the algorithm's computation is bounded by

$$\sum_i (r_i + deg_i - 1) \cdot k^{w^*+1}.$$

We can argue that $\sum_i deg_i \leq n$, when n is the number of variables, because only a single function is generated in each bucket, and there are total of n buckets. Therefore, the total complexity can be bound by $O((r + n) \cdot k^{w^*+1})$. Assuming $r > n$, this becomes $O(r \cdot k^{w^*+1})$. The size of each λ message is $O(k^{w^*})$. Since the total number of λ messages is bounded by n, the total space complexity is $O(n \cdot k^{w^*})$. □

CHAPTER 5

Tree-Clustering Schemes

In this chapter, we take the bucket elimination algorithm a step further. We will show that bucket elimination can be viewed as an algorithm that sends messages along a tree (the bucket tree). The algorithm can then be augmented with a second set of messages passed from bottom to top, yielding a message-passing schemes that belongs to the class of *cluster tree elimination* algorithms.

These latter methods have received different names in different research areas, such as join-tree clustering or junction-tree algorithms, clique-tree clustering, and hyper-tree decompositions. We will refer to all these as cluster-tree processing schemes over *tree-decompositions*. Our algorithms are applicable to a general reasoning problem described by $\mathcal{P} = \langle \mathbf{X}, \mathbf{D}, \mathbf{F}, \bigotimes, \Downarrow \rangle$, where the first four elements identify the graphical model and the fifth identifies the reasoning task (see Definition 2.2).

Important: we will assume the specific of probabilistic networks when developing the algorithms and the reader can just make the appropriate generalization. Also we will allow abuse of notation when including in a model its query operator whenever relevant. Henceforth we assume $\mathcal{M} = \langle \mathbf{X}, \mathbf{D}, \mathbf{F}, \prod, \sum \rangle$.

5.1 BUCKET-TREE ELIMINATION

The bucket-elimination algorithm, *BE-bel* (see Figure 4.5) for belief updating is designed to compute the belief of the first node in a given ordering, and the probability of evidence. However, it is often desirable to answer the belief query for each and every variable in the network. A brute-force approach will require running *BE-bel* n times, each time with a different variable at the start of the ordering. We will show next that this is unnecessary. By viewing bucket-elimination as a message passing algorithm along a rooted *bucket tree*, we can augment it with a second message passing phase in the opposite direction, from root to leaves, achieving the same goal.

Example 5.1 Consider our ongoing Bayesian network example defined over the directed acyclic graph (DAG) in Figure 4.1 and appearing again here in Figure 5.8(a). Figure 5.1a recaps the initial buckets along ordering $d = (A, B, C, D, F, G)$ and the messages, labeled by λ, that will be passed by BE from top to bottom. Figure 5.1b depicts the same computation as message-passing along a tree which we will refer to as a *bucket tree*. Notice that, as before, the ordering is displayed from the bottom up (A, the first variable, is at the bottom and G, the last one, is at the top), and the messages are passed top down. This computation results in the belief in A, $bel(A) = P(A|G = 1)$, and consulted only functions that reside in the bucket of A. What if we now want to compute $bel(D)$? We can start the algorithm using a new ordering such as (D, A, B, C, F, G). Alternatively, rather then doing all the computations from scratch using a different variable ordering whose first variable is D, we can take the bucket tree and re-orient the edges to make D the root of the tree. Reorienting the tree so that D is the root, requires reversing only 2 edges, (B, D) and (A, B), suggesting that we only need to recompute messages from node A to B and from B to D. We can think about a new virtual partial order expressed as $(\{D, A, B\}, C, F, G)$, namely, collapsing the buckets of B, A and D into a single bucket and therefore ignoring their internal order. By definition, we can compute the belief in D by the expression

$$bel(d) = \alpha \sum_a \sum_b P(a) \cdot p(b|a) \cdot P(d|a, b) \cdot \lambda_{C \to B}(b) . \tag{5.1}$$

Likewise, we can also compute the belief in B by

$$bel(b) = \alpha \sum_a \sum_d P(a) \cdot p(b|a) \cdot P(d|a, b) \cdot \lambda_{C \to B}(b) . \tag{5.2}$$

This computation can be carried also over the bucket tree, whose downward messages were already passed, in three steps. The first executed in bucket A, where the function $P(A)$ is moved to $bucket_B$, the second is executed by $bucket_B$, computing a function (a product) that is moved to $bucket_D$. The final computation is carried in $bucket_D$. Denoting the new reverse messages by π, a new $\pi_{A \to B}(a) = P(A)$, is passed from $bucket_A$ to $bucket_B$. Then, an intermediate function is computed in $bucket_B$, to be sent to $bucket_D$, using the messages received from $bucket_C$ and $bucket_A$ and its own function,

$$\pi_{B \to D}(a, b) = p(b|a) \cdot \pi_{A \to B}(a) \cdot \lambda_{C \to B}(b) .$$

Finally, the belief is computed in $bucket_D$ using its current function content by

$$bel(d) = \alpha \sum_{a,b} P(d|a, b) \cdot \pi_{B \to D}(a, b). \tag{5.3}$$

This accomplishes the computation of the algebraic expression in Eq (5.1). You can see some of these messages depicted in Figure 5.2a. The belief in B can also be computed in $bucket_D$. However, if we want each bucket to compute its own belief, and since $bucket_D$ sends $P(D|A, B)$ to $bucket_B$ anyway, as part of BE, the computation of Eq. (5.2) can be carried out there, autonomously.

The example generalizes. We can compute the belief for every variable by a second message passing from the root to the leaves along the bucket tree, such that at termination the belief for each variable can be computed locally, in each bucket, consulting only the functions in its own bucket.

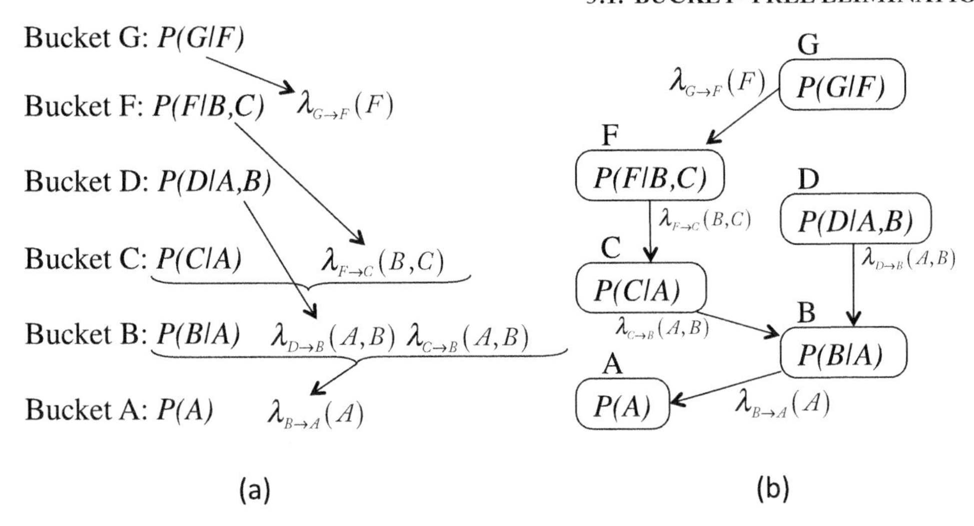

Figure 5.1: Execution of BE along the bucket tree.

Let $\mathcal{M}$ be a graphical model $\mathcal{M} = \langle \mathbf{X}, \mathbf{D}, \mathbf{F}, \prod \rangle$ and d an ordering of its variables $X_1, ..., X_n$. Let $B_{X_1}, ..., B_{X_n}$ denote a set of buckets, one for each variable. We will use B_{X_i} and B_i interchangeably. As before, each bucket B_i contains those functions in F whose latest variable in ordering d is X_i (i.e., according to the bucket-partitioning rule), and, as before, we denote by ψ_i the product of functions in B_i. A bucket tree of a model $\mathcal{M}$ along d has buckets as its nodes. Bucket B_X is connected to bucket B_Y if the function generated in bucket B_X by BE is placed in B_Y. Therefore, in a bucket tree, every vertex B_X other than the root has one parent vertex B_Y and possibly several child vertices $B_{Z_1}, ..., B_{Z_t}$.

The structure of the bucket tree can be extracted also from the induced-ordered graph of $\mathcal{M}$ along d using the following definition.

Definition 5.2 bucket-tree, separator, eliminator. Let $\mathcal{M} = \langle \mathbf{X}, \mathbf{D}, \mathbf{F}, \prod \rangle$ be a graphical model whose primal graph is G, and let $d = (X_1, ..., X_n)$ be an ordering of its variables. Let (G^*, d) be the induced graph along d of G.

- The bucket tree has the buckets denoted $\{B_i\}_{i=1}^n$ as its nodes. Each bucket contains a set of functions and a set of variables. The functions are those placed in the bucket according to the bucket partitioning rule where ψ_i is their product. The set of variables in B_i, denoted $scope(B_i)$, is X_i and all its parents in the induced-graph (G^*, d). Each vertex B_i points to B_j (or, B_j is the parent of B_i) if X_j is the closest neighbor of X_i that appear before it in (G^*, d).

- If B_j is the parent of B_i in the bucket tree, then the separator of X_i and X_j, $sep(B_i, B_j) = scope(B_i) \cap scope(B_j)$.

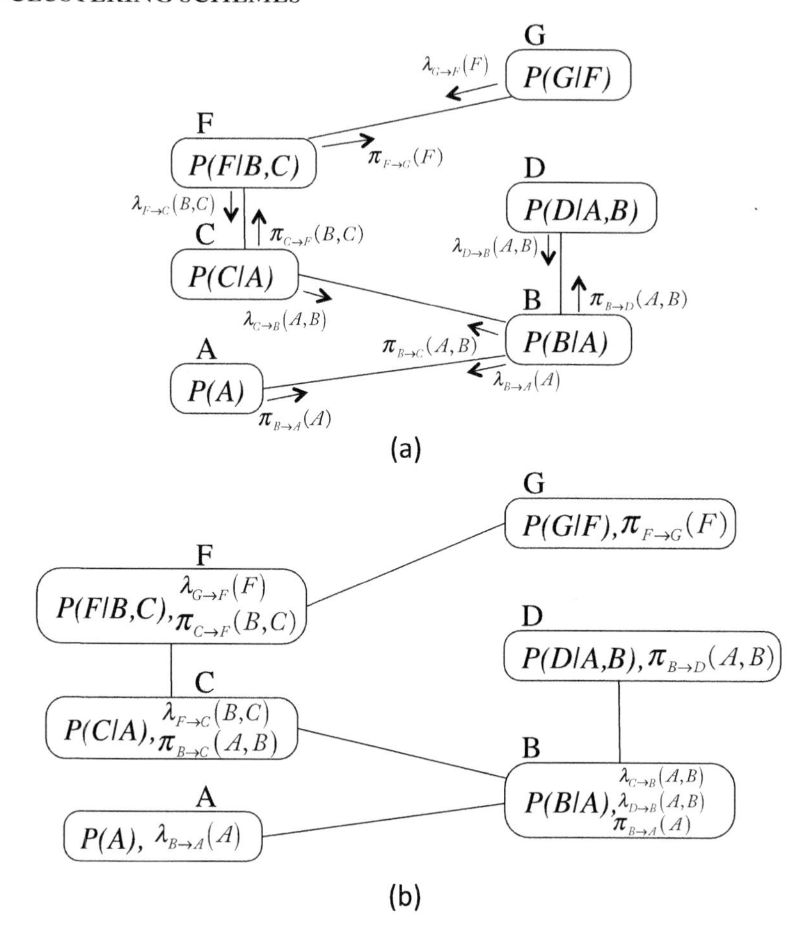

Figure 5.2: Propagation of π's and λ's along the bucket tree (a), and the augmented output bucket tree (b).

- Given a directed edge (B_i, B_j) in the bucket tree, $elim(i, j)$ is the set of variables in B_i and not in B_j, namely $elim(B_i, B_j) = scope(B_i) - sep(B_i, B_j)$. We will call this set "the eliminator from B_i to B_j."

Algorithm *bucket-tree elimination (BTE)* presented in Figure 5.3 includes the two message passing phases along the bucket tree. Notice that we always assume that the input may contain also a set of evidence nodes because this is typical to a large class of problems in probabilistic networks. Given an ordering of the variables, the first step of the algorithm generates the bucket tree by partitioning the functions into buckets and connecting the buckets into a tree. The subsequent *top-down* phase is identical to general bucket elimination. The *bottom-up* messages are defined as follows. The messages sent from the root up to the leaves will be denoted by π. The message from B_j to a child B_i is generated by multiplying the bucket's function ψ_j by all the π messages from its parent bucket and all the λ messages from its *other* child buckets and marginalizing (e.g., summing) over the eliminator from B_j to B_i. We see that downward messages are generated by eliminating a single variable. Upward messages, on the other hand, may be generated by eliminating zero, one or more variables.

When BTE terminates, each output bucket B'_i contains the $\pi_{j\to i}$ it received from its parent B_j, its own function ψ_j and the $\lambda_{k\to i}$ messages sent from each child B_k. Then, each bucket can compute its belief over all the variables in its bucket, by multiplying all the functions in a bucket as specified in step 3 of BTE. It can then compute also the belief on single variables and the probability of evidence as in the procedure in Figure 5.4.

ALGORITHM BUCKET-TREE ELIMINATION (BTE)

Input: A problem $\mathcal{M} = \langle \mathbf{X}, \mathbf{D}, \mathbf{F}, \prod, \sum \rangle$, ordering d.
$X = \{X_1, ..., X_n\}$ and $F = \{f_1, ..., f_r\}$
Evidence $E = e$.

Output: Augmented buckets $\{B'_i\}$, containing the original functions and all the π and λ functions received from neighbors in the bucket tree.

1. **Pre-processing:** Partition functions to the ordered buckets as usual and generate the bucket tree.
2. **Top-down phase:** λ messages (BE) **do**
   for $i = n$ to 1, in reverse order of d process bucket B_i:
   The message $\lambda_{i\to j}$ from B_i to its parent B_j, is:
   $\lambda_{i\to j} \Leftarrow \sum_{elim(i,j)} \psi_i \cdot \prod_{k\in child(i)} \lambda_{k\to i}$
   endfor
3. **bottom-up phase: π messages**
   for $j = 1$ to n, process bucket B_j **do**:
   B_j takes $\pi_{k\to j}$ received from its parent B_k, and computes a message $\pi_{j\to i}$ for each child bucket B_i by
   $\pi_{j\to i} \Leftarrow \sum_{elim(j,i)} \pi_{k\to j} \cdot \psi_j \cdot \prod_{r\neq i} \lambda_{r\to j}$
   endfor
4. **Output:** augmented buckets $B'_1, ..., B'_n$, where each B'_i contains the original bucket functions and the λ and π messages it received.

Figure 5.3: Algorithm bucket-tree elimination.

COMPUTING MARGINAL BELIEFS

Input: a bucket tree processed by BTE with augmented buckets: $B'_1, \ldots, B'_n$
output: beliefs of each variable, bucket, and probability of evidence.

$bel(B_i) \Leftarrow \prod_{f \in B'_i} f$

$bel(X_i) \Leftarrow \sum_{B_i - \{X_i\}} \prod_{f \in B'_i} f$

$P(evidence) \Leftarrow \sum_{B_i} \prod_{f \in B'_i} f$

Figure 5.4: Query answering.

Example 5.3 Figure 5.2a shows the complete execution of BTE along the bucket tree. Notice, that the variables in each bucket are not stated explicitly in the figure. For example, the variables in the $bucket_C$ are A, B, C while the scope of the original function has only variables A, C. The π and λ messages are placed on the outgoing upward arcs. The π functions in the bottom-up phase are computed as follows (the first three were demonstrated earlier):
$\pi_{A \to B}(a) = P(a)$
$\pi_{B \to C}(c, a) = P(b|a)\lambda_{D \to B}(a, b)\pi_{A \to B}(a)$
$\pi_{B \to D}(a, b) = P(b|a)\lambda_{C \to B}(a, b)\pi_{A \to B}(a, b)$
$\pi_{C \to F}(c, b) = \sum_a P(c|a)\pi_{B \to C}(a, b)$
$\pi_{F \to G}(f) = \sum_{b,c} P(f|b, c)\pi_{C \to F}(c, b)$

The actual output (the augmented buckets) are shown in Figure 5.2b.

Explicit submodels. Extending the view of the above algorithms for any reasoning task over graphical models, we can show that when BTE terminates we have in each bucket all the information needed to answer any reasoning task on the variables appearing in that bucket. In particular, we do not need to look outside a bucket to answer a belief query. We call this property "explicitness". It is sometime referred to also as *minimality* or *decomposability* [Montanari, 1974].

Definition 5.4 Explicit function and explicit sub-model. Given a graphical model $\mathcal{M} = \langle \mathbf{X}, \mathbf{D}, \mathbf{F}, \prod \rangle$, and reasoning tasks defined by marginalization $\sum$ and given a subset of variables $Y, Y \subseteq \mathbf{X}$, we define $\mathcal{M}_Y$, the explicit function of $\mathcal{M}$ over Y:

$$\mathcal{M}_Y = \sum_{X-Y} \prod_{f \in F} f, \tag{5.4}$$

We denote by F_Y any set of functions whose scopes are subsumed in Y over the same domains and ranges as the functions in $\mathbf{F}$. We say that (Y, F_Y) is an explicit submodel of $\mathcal{M}$ iff

$$\prod_{f \in F_Y} f = \mathcal{M}_Y \tag{5.5}$$

As we elaborate more later, once we have an explicit representation in each cluster we can answer most queries locally. We can compute the belief of each variable, the probability of evidence or the partition function, and the optimal solution costs inside each bucket.

Theorem 5.5 Completeness of BTE. *Given $\mathcal{M} = \langle \mathbf{X}, \mathbf{D}, \mathbf{F}, \prod, \sum \rangle$ and evidence $E = e$, when algorithm BTE terminates, each output bucket is explicit relative to its variables. Namely, for each B_i, $\Pi_{f \in B'_i} f = \mathcal{M}_{scope(B_i)}$.*

The completeness of BTE will be derived from that of a larger class of tree propagation algorithms that we present in the following section. We next address the complexity of BTE.

Theorem 5.6 Complexity of BTE. *Let $w^*(d)$ be the induced width of (G^*, d) where G is the primal graph of $\mathcal{M} = \langle \mathbf{X}, \mathbf{D}, \mathbf{F}, \prod, \sum \rangle$, r be the number of functions in $\mathbf{F}$ and k be the maximum domain size. The time complexity of BTE is $O(r \cdot deg \cdot k^{w^*(d)+1})$, where deg is the maximum degree of a node in the bucket tree. The space complexity of BTE is $O(n \cdot k^{w^*(d)})$.*

Proof: As we previously showed, the downward λ messages take $O(r \cdot k^{w^*(d)+1})$ steps. This simply is the complexity of BE. The upward messages per bucket are computed for each of its child nodes. Each such message takes $O(r_i \cdot k^{w^*(d)+1})$ steps, where r_i is the number of functions in bucket B_i, yielding a time complexity per upward bucket of $O(r_i \cdot deg \cdot k^{w*d)+1})$. Summing over all buckets we get complexity of $O(r \cdot deg \cdot k^{w^*(d)+1})$. Since the size of each downward message is k^{w^*} we get space complexity of $O(n \cdot k^{w^*(d)})$. The complexity of BTE can be improved to be $O(rk^{w^*(d)+1})$ time and $O(nk^{w^*(d)+1})$ space [Kask and Dechter, 2005]. □.

In theory the speedup expected from running BTE vs. running BE n times is at most n. This may seem insignificant compared with the exponential complexity in w^*, however it can be very significant in practice, especially when n is large. Beyond the saving in computation, the bucket tree provides an architecture for distributed computation of the algorithm when each bucket is implemented by a different cpu.

5.1.1 ASYNCHRONOUS BUCKET-TREE PROPAGATION

Algorithm *BTE* can also be described without committing to a particular schedule by viewing the bucket tree as an undirected tree and by unifying the up and down messages into a single message-type denoted by λ. In this case each bucket receives a λ message from each of its neighbors and each sends a λ message to every neighbor. This distributed algorithm, called Bucket-Tree

Propagation or *BTP*, is written for a single bucket described in Figure 5.5. It is easy to see that the algorithm accomplishes the same as BTE and is therefore correct and complete. It sends at most two messages on each edge in the tree (computation starts from the leaves of the tree). It is also easy to see the distributed nature of the algorithm.

Theorem 5.7 Completeness of BTP. *Algorithm BTP terminates generating explicit buckets.*

Proof. The proof of BTP correctness follows from the correctness of BTE. All we need to show is that at termination the buckets' content in BTE and BTP are the same. (Prove as an exercise.) □

BUCKET-TREE PROPAGATION (BTP)

Input: A problem $\mathcal{M} = \langle \mathbf{X}, \mathbf{D}, \mathbf{F}, \prod, \sum \rangle$, ordering d. $X = \{X_1, ..., X_n\}$ and $F = \{f_1, ..., f_r\}$, $\mathbf{E} = \mathbf{e}$. An ordering d and a corresponding bucket-tree structure, in which for each node X_i, its bucket B_i and its neighboring buckets are well defined.

Output: Explicit buckets. Assume functions assigned with the evidence.

1. **for** bucket B_i **do:**
2. **for** each neighbor bucket B_j **do,**
   once all messages from all other neighbors were received, **do**
   compute and send to B_j the message
   $\lambda_{i \to j} \Leftarrow \sum_{elim(i,j)} \psi_i \cdot (\prod_{k \neq j} \lambda_{k \to i})$
3. **Output:** augmented buckets $B'_1, ..., B'_n$, where each B'_i contains the original bucket functions and the λ messages it received.

Figure 5.5: Algorithm Bucket-tree propagation (BTP).

Finally, since graphical models whose primal graph is a tree, the induced-width is 1, clearly

Proposition 5.8 BTE on trees *For tree graphical models, algorithms BTE and BTP are time and space $O(nk^2)$ and $O(nk)$, respectively, when k bound the domain size and n bounds the number of variables.*

5.2 FROM BUCKET TREES TO CLUSTER TREES

Algorithms BTE and its synchronous version BTP are special cases of a class of algorithms that operates over a tree-decomposition of the graphical model. A tree-decomposition takes a graphical model and embeds it in a tree where each node in the tree is a cluster of variables and functions. The decomposition allows message-passing between the clusters in a manner similar to BTE and BTP, yielding an algorithm which we will call Cluster-Tree Elimination or CTE. We provide the idea through a short route from BTE to CTE and subsequently establish the formal grounds in more details.

5.2.1 FROM BUCKETS TO CLUSTERS; THE SHORT ROUTE

Example 5.9 Let's go back to example 5.1. We saw that to facilitate message-passing in the reverse order of d we suggested first to virtually collapse the individual buckets of D, A, B into a single cluster and reasoned from there. However, we can actually apply this collapsing and then perform the computation defined in Equations 5.1 and 5.2 within the collapsed cluster, instead of moving around information in between the individual buckets. This will yield 4 clusters. The first 3 correspond to the original buckets of variables G, F and C, and the forth is the cluster over $\{A, D, B\}$ groups all the functions in the respective buckets. The message sent from the new cluster ABD to $bucket_C$ is identical to the π message from $bucket_b$ to $bucket_C$, and it can be computed using the the same rule as specified in algorithm BTP. Namely, the message from ABD to C is computed by taking the product of all the functions in cluster ABD and then eliminating the eliminator (namely, summing over D). We get,

$$\pi_{ABD \to C} = \sum_D P(b|a) \cdot P(d|a, b) \cdot P(a).$$

It is easy to see that this is the same $\pi_{B \to C}$ computed by BTP.

In other words, BTE/BTP algorithms can be extended to work over a larger ensemble of cluster-trees and those can be obtained, by just collapsing some adjacent clusters into larger clusters, where the starting point is a bucket-tree. In certain cases, as in the example above, there is no loss in efficiency. In fact we have less clusters yielding a simplification. In other cases, if the clusters get larger and larger, we may loose decomposability and therefore have less effective computation. We will identify the tradeoffs associated with this process and provide a scheme to generate good cluster-tree decompositions. What we wish to stress here is that the only thing that will change is the complexity of the resulting algorithms.

In summary, the algorithm CTE that we will present, is just the BTP that is applied to any cluster-tree and those can be obtained by collapsing adjacent buckets (their functions and variables) of a bucket-trees. A preliminary version of the algorithm, called *CTP* is presented in Figure 5.6.

CLUSTER-TREE PROPAGATION (CTP)

Input: A problem $\mathcal{M} = \langle \mathbf{X}, \mathbf{D}, \mathbf{F}, \prod, \sum \rangle$, ordering d. $X = \{X_1, ..., X_n\}$ and $F = \{f_1, ..., f_r\}$, $\mathbf{E} = \mathbf{e}$. An ordering d and a corresponding bucket-tree structure, in which for each node X_i, its bucket B_i and its neighboring buckets are well defined.

Output: Explicit buckets.

0. $CT \leftarrow$ Generate clusters $C_1, ...C_l$ collapsing, some adjacent buckets.
1. **for** bucket C_i **do:**
2. **for** each neighbor bucket C_j **do**,
   once all messages from all other neighbors were received, **do**
   compute and send to C_j the message
   $\lambda_{i \to j} \Leftarrow \sum_{elim(i,j)} \psi_i \cdot (\prod_{k \neq j} \lambda_{k \to i})$
3. **Output:** augmented buckets $C'_1, ..., C'_n$, where each C'_i contains the original bucket functions and the λ messages it received.

Figure 5.6: Algorithm Cluster-Tree propagation (CTP).

In the rest of this chapter we provide a somewhat different route to CTE and to general message-passing over general tree-decompositions, whose starting point are acyclic graphical models. Some graphical models are inherently tree-structured. Namely, their input functions already has a dependency structure that can be captured by a tree-like graph (with no cycles). Such graphical models are called *Acyclic Graphical models* [Maier, 1983] and they include regular trees as a special case. We will describe acyclic models first and then show how a tree-decomposition can impose a tree structure on non-acyclic graphical models as well. In particular we will show that bucket-trees are special cases of tree-decompositions.

5.2.2 ACYCLIC GRAPHICAL MODELS

As we know, a graphical model can be associated with a dual graph (see Definition 2.7). which provides an alternative view of the graphical model. In this view, each function resides in its own node which can be regarded as a meta variabls and the arcs indicate equality constraints between shared variables. So, if a graphical model's dual graph is a tree, we have a tree-graphical model, and we know it can be solved in linear time by a BTE-like message-passing algorithm over the dual graph (see Proposition 5.8.)

Sometime the dual graph seems to not be a tree, but it is in fact, a tree. This is because some of its arcs are redundant and can be removed while not violating the original independency relationships that is captured by the graph. Arcs are redundant if they express a dependency between two nodes that is already captured or implied by an alternative set of dependencies, and therefore

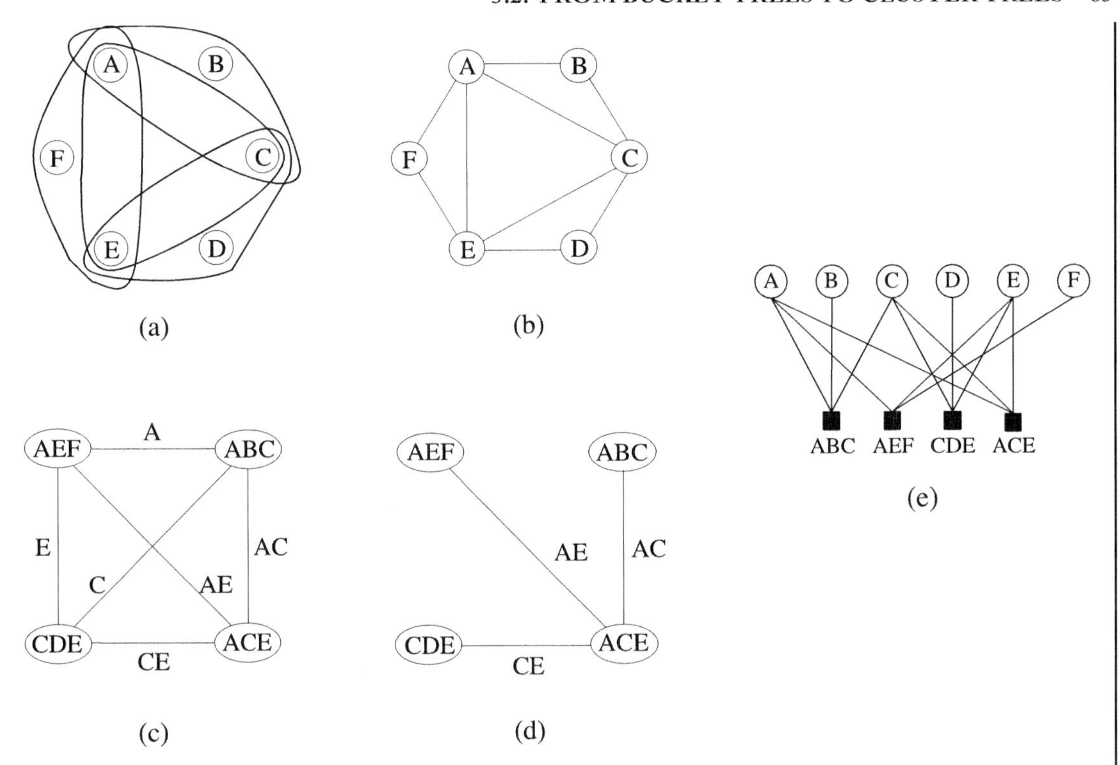

Figure 5.7: (a) Hyper; (b) primal; (c) dual; (d) join-tree of a graphical model having scopes ABC, AEF, CDE and ACE; and (e) the factor graph.

their removal does not alter the conditional independence captured by graph separation (for more see [Pearl, 1988]). We illustrate with the next example and then provide the formal definition of capturing this notion.

Example 5.10 Refer to Figure 5.7 (presented earlier Figure 2.1). We see that the arc between (AEF) and (ABC) in Figure 2.1c expresses redundant dependency because variable A also appears along the alternative path $(ABC) - AC - (ACE) - AE - (AEF)$. In other words, a dependency between AEF and ABC relative to A is maintained through the path even if they are not directly connected. Likewise, the arcs labeled E and C are also redundant. Their removal yields the tree in 2.1d which we call a join-tree. This reduced dual graph in 2.1d satisfies a property called connectedness.

Definition 5.11 Connectedness, join-trees. Given a dual graph of a graphical model $\mathcal{M}$, an arc subgraph of the dual graph satisfies the *connectedness* property iff for each two nodes that share a variable, there is at least one path of labeled arcs of the dual graph such that each contains the shared variables. An arc subgraph of the dual graph that satisfies the connectedness property is called a *join-graph* and if it is a tree, it is called a *join-tree.*

We can now formally define acyclic graphical models.

Definition 5.12 Acyclic networks. A graphical model , $\mathcal{M} = \langle X, D, F, \prod \rangle$, whose dual graph *has* a join-tree is called an *acyclic graphical model.*

Example 5.13 We can see that the join-tree in Figure 5.7d satisfies the connectedness property. The graphical model defined by the scopes: AEF, ABC, CDE, ACE is therefore, acyclic.

It is easy to see that if algorithm BTE is applied to an acyclic graphical model it would be efficient. In fact, it will be time and space linear. This is because the join-tree suggests an ordering over the variables where the width of the primal graph along d equals the induced-width (prove as an exercise) and where each function is placed in a single bucket. This means that messages are always defined over scopes that are subsumed by existing scopes of the original functions.

Theorem 5.14 *Given an acyclic graphical model, algorithm BTE is time and space linear. (See proof in the Appendix.)*

5.2.3 TREE DECOMPOSITION AND CLUSTER TREE ELIMINATION

Now that we have established that acyclic graphical models can be solved efficiently, all that remains is to transform a general graphical model into an acyclic one. This task is facilitated by algorithm tree-decomposition defined next. We will subsequently show that a bucket tree is a special case of tree decomposition.

Definition 5.15 Tree decomposition, cluster tree. Let $\mathcal{M} =< X, D, F, \prod >$ be a graphical model. A *tree-decomposition* of $\mathcal{M}$ is a triple $< T, \chi, \psi >$, where $T = (V, E)$ is a tree, and χ and ψ are labeling functions which associate with each vertex $v \in V$ two sets, $\chi(v) \subseteq X$ and $\psi(v) \subseteq F$ satisfying:

1. for each function $f_i \in F$, there is *exactly* one vertex $v \in V$ such that $f_i \in \psi(v)$, and $scope(f_i) \subseteq \chi(v)$; and
2. for each variable $X_i \in X$, the set $\{v \in V | X_i \in \chi(v)\}$ induces a connected subtree of T. This is also called the running intersection property [Maier, 1983].

We will often refer to a node and its functions as a *cluster* and use the term *tree decomposition* and *cluster tree* interchangeably.

Definition 5.16 treewidth, pathwidth, separator-width, eliminator. The *treewidth* [Arnborg, 1985] of a tree decomposition $< T, \chi, \psi >$ is $max_{v \in V} |\chi(v)|$ minus 1. Given two adjacent vertices u and v of a tree-decomposition, the *separator* of u and v is $sep(u, v) = \chi(u) \cap \chi(v)$, and the *eliminator* of u with respect to v is $elim(u, v) = \chi(u) - \chi(v)$. The separator-width is the maximum over all separators. The pathwidth of a graph is the treewidth when only chain-line tree-decompositions are restricted.

Example 5.17 Consider again the Bayesian network in Figure 5.8a. Any of the cluster-trees in Figure 5.9 describes a partition of variables into clusters. We can now place each input function into a cluster that contains its scopes, and verify that each is a legitimate tree decomposition. For example, Figure 5.9c shows a cluster-tree decomposition with two vertices, and labeling $\chi(1) = \{G, F\}$ and $\chi(2) = \{A, B, C, D, F\}$. Any function with scope $\{G\}$ must be placed in vertex 1 because vertex 1 is the only vertex that contains variable G (placing a function having G in its scope in another vertex will force us to add variable G to that vertex as well). Any function with scope $\{A, B, C, D\}$ or one of its subsets must be placed in vertex 2, and any function with scope $\{F\}$ can be placed either in vertex 1 or 2. Notice that the tree-decomposition at Figure 5.9a is actually a bucket-tree.

We see that for some nodes $sep(u, v) = \chi(u)$. That is, all the variables in vertex u belong to an adjacent vertex v. In this case the number of clusters in the tree decomposition can be reduced by merging vertex u into v without increasing the cluster size in the tree-decomposition. This is accomplished by moving from Figure 5.9a to Figure 5.9b.

Definition 5.18 Minimal tree decomposition. A tree decomposition is *minimal* if $sep(u, v) \subsetneqq \chi(u)$ and $sep(u, v) \subsetneqq \chi(v)$ for each pair (u, v) of adjacent nodes.

We can show the following.

Theorem 5.19 *A bucket tree of a graphical model $\mathcal{M}$, is a tree decomposition of $\mathcal{M}$. (for a proof see Appendix.)*

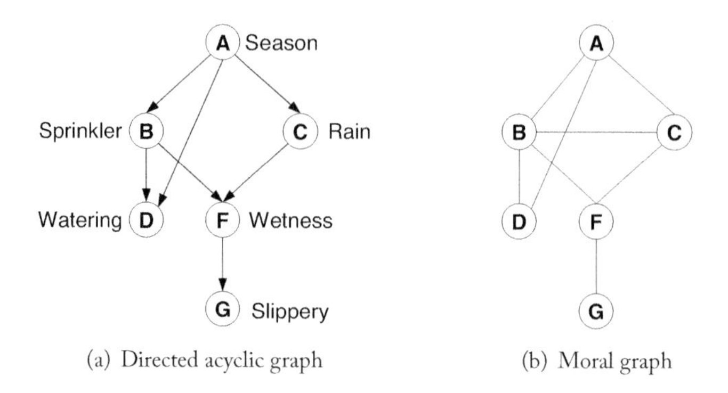

(a) Directed acyclic graph (b) Moral graph

Figure 5.8: Belief network $P(G, F, C, B, A) = P(G|F)P(F|C, B)P(D|A, B)P(C|A)P(B|A)P(A)$.

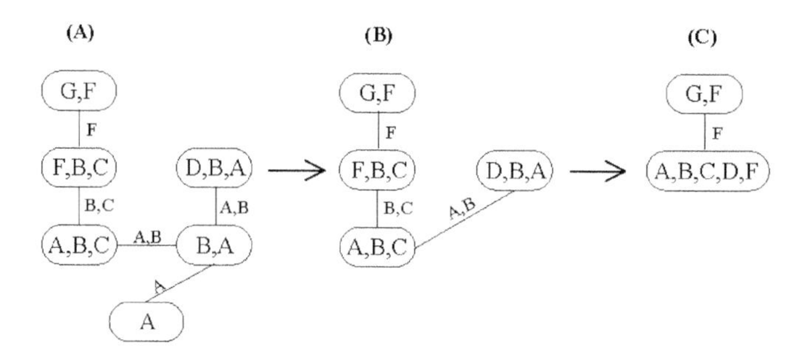

Figure 5.9: From a bucket tree to join tree to a super bucket tree.

CTE. We will show now that a tree decomposition facilitates a message-passing scheme, called *Cluster-Tree Elimination (CTE)*, that is similar to BTE in the same sense the CTP is similar to BTP. The algorithm is presented in Figure 5.10. Like in BTP, each vertex of the tree sends a function to each of its neighbors. All the functions in a vertex u and all the messages received by u from all its neighbors other than a specific vertex v to which u's message is directed, are combined by product. The combined function is marginalized over the separator of vertices u and v (namely, eliminating the eliminator) using the marginalization operator, $\sum$, and the marginalized function is then sent from u to v. We will denote message by m.

Vertex activation can be asynchronous and convergence is guaranteed. If processing is performed from leaves to root and back, convergence is guaranteed after two passes, where only one message is sent on each edge in each direction. If the tree contains l edges, then a total of $2l$ messages will be sent.

Example 5.20 Consider again the graphical model whose primal graph appears in Figure 5.8(b) but now assume that all functions are defined on pairs of variables (you can think of this as a

CLUSTER-TREE ELIMINATION (CTE)

Input: A tree decomposition $< T, \chi, \psi >$ for a problem $M =< X, D, F, \prod, \sum\} >$, $X = \{X_1, ..., X_n\}$, $F = \{f_1, ..., f_r\}$. Evidence $E = e$, $\psi_u = \prod_{f \in \psi(u)} f$

Output: An augmented tree decomposition whose clusters are all model explicit. Namely, a decomposition $< T, \chi, \bar{\psi} >$ where $u \in T$, $\bar{\psi}(u)$ is model explicit relative to $\chi(u)$.

1. **Initialize.** (denote by $m_{u \to v}$ the message sent from vertex u to vertex v.)
2. **Compute messages:**

   For every node u in T, once u received messages from all neighbors but v,

   Process observed variables:

   For each node $u \in T$ assign relevant evidence to $\psi(u)$

   Compute the message:

   $$m_{u \to v} \leftarrow \sum_{\chi(u)-sep(u,v)} \psi_u \cdot \prod_{r \in neighbor(u), r \neq v} m_{r \to u}$$

   endfor

   Note: functions whose scopes do not contain any separator variable do not need to be combined and can be directly passed on to the receiving vertex.
3. **Return:** The explicit tree $< T, \chi, \bar{\psi} >$, where

   $\bar{\psi}(v) \Leftarrow \psi(v) \cup_{u \in neighbor(v)} \{m_{u \to v}\}$

   return the explicit function: for each v, $M_{\chi(v)} = \prod_{f \in \bar{\psi}(v)} f$

Figure 5.10: Algorithm Cluster-Tree Elimination (CTE).

Markov network). Two tree decompositions are given in Figure 5.11a and 5.11b. For the tree-decomposition in 5.11b we show the propagated messages explicitly in Figure 5.11c. Since cluster 1 contains only one function, the message from cluster 1 to 2 is the summation of f_{FG} over the separator between cluster 1 and 2, which is variable F. The message $m_{2 \to 3}$ from cluster 2 to cluster 3 combines the functions in cluster 2 with the message $m_{1 \to 2}$, and marginalizes over the separator between cluster 2 and 3, yielding $\{B, C\}$, and so on.

Once all vertices have received messages from all their neighbors we have the explicit clusters (see Definition 5.4) and therefore an answer to any singleton marginal query (e.g., beliefs) and a host of other reasoning tasks can be accomplished over the explicit tree in linear time. Before we prove these properties in Section 5.3 we will pause to discuss the generation of tree-decompositions.

5.2.4 GENERATING TREE DECOMPOSITIONS

We have already established that a bucket-tree built along a given ordering is a tree decomposition. Each node is a bucket, whose variables include the bucket's variable and all its earlier

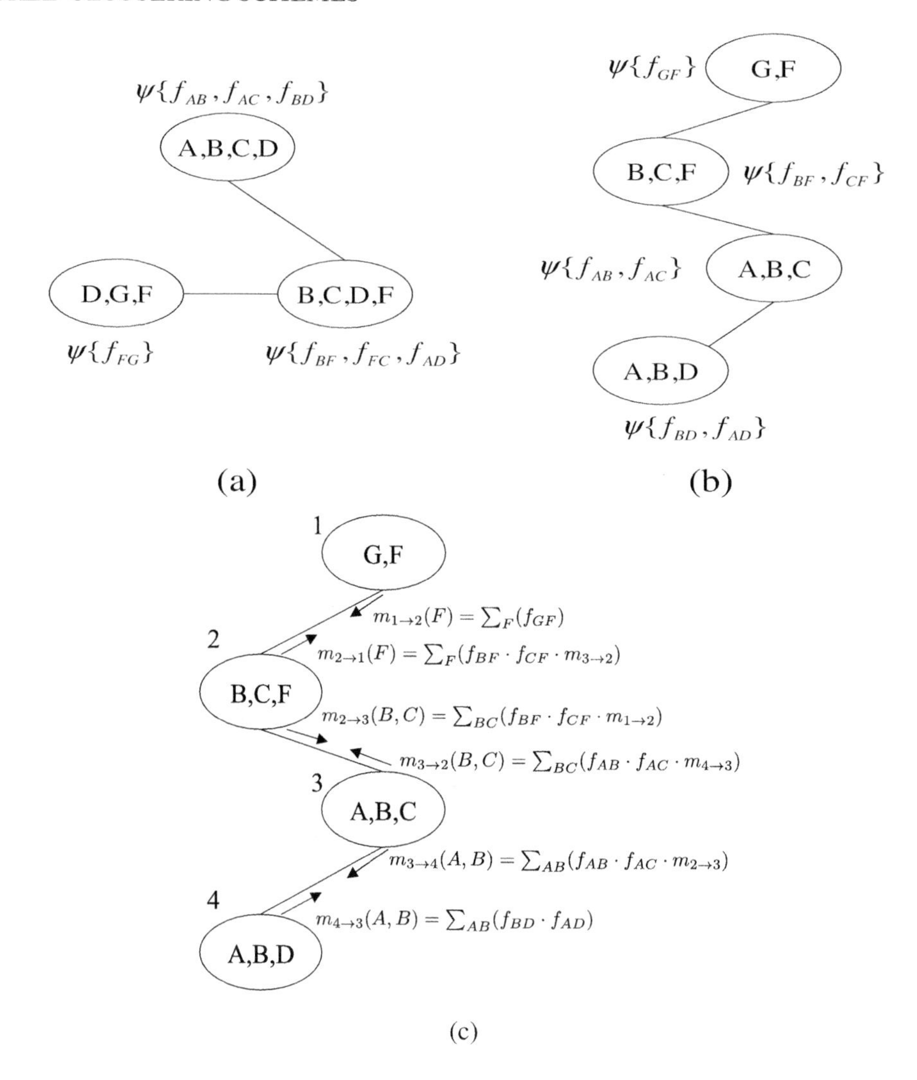

Figure 5.11: Two tree decompositions of a graphical model.

neighbors in the induced graph and whose functions are those assigned to it by the initial bucket-partitioning. As suggested earlier, once we have a tree decomposition, other tree decompositions can be obtained by merging adjacent nodes. Therefore, bucket trees can serve as a starting point for generating arbitrary tree decompositions, a process that is justified by the following proposition.

Proposition 5.21 *If T is a tree decomposition, then any tree obtained by merging adjacent clusters is also a tree decomposition. (Prove as an exercise.)*

A special class of tree-decompositions called *join-trees* can be obtained by merging subsumed buckets into their containing buckets, as defined by their variables. This is indeed the approach we discussed in our short route. Alternatively, they can be generated directly from the induced-graph by selecting as clusters only those buckets that are associated with maximal cliques.

Algorithm join-tree clustering (JTC) for generating join-tree decompositions, is described in Figure 5.12. The algorithm generates an induced graph, which we know is chordal, it then identifies its maximal cliques as the candidate cluster-nodes and then connect them in a tree structure. This process determines the variables associated with each node. Subsequently, functions are partitioned into the clusters appropriately.

JOIN-TREE CLUSTERING (JTC)
Input: A graphical model $\mathcal{M} = \langle \mathbf{X}, \mathbf{D}, \mathbf{F}, \prod \rangle$, $\mathbf{X} = \{X_1, ..., X_n\}$, $\mathbf{F} = \{f_1, ..., f_r\}$.
Its scopes $S = S_1, ..., S_r$ and its primal graph is $G = (X, E)$.
Output: A join-tree decomposition $< T, \chi, \psi >$ for $\mathcal{M}$
1. Select a variable ordering, $d = (X_1, ..., X_n)$.
2. **Triangulation** (create the induced graph along d and call it G^*):
    for $j = n$ to 1 by -1 do
    $E \leftarrow E \cup \{(i, k) | i < j, \ k < j, \ (i, j) \in E, (k, j) \in E\}$
3. **Create a join-tree of the induced graph** (G^*, d) as follows:
    a. Identify all maximal cliques in the chordal graph.
    Let $C = \{C_1, ..., C_t\}$ be all such cliques, where C_i is the cluster of bucket i.
    b. Create a tree T of cliques:
    Connect each C_i to a C_j $(j < i)$ with whom it shares largest subset of variables.
4. Create ψ_i: Partition input function in cluster-node whose variables contain its scope.
5. Return a tree-decomposition $< T, \chi, \psi >$, where T is generated in step 3,
$\chi(i) = C_i$ and $\psi(i)$ is determined in step 4.

Figure 5.12: Join-tree clustering.

Example 5.22 Consider the graph in Figure 5.13a and the ordering $d_1 = (G, D, F, C, B, A)$ in Figure 5.13b. Performing the triangulation step of JTC connects parents recursively from the last variable to the first, creating the induced-ordered graph by adding the new (broken) edges of Figure 5.13b. The maximal cliques of this induced graph are: $Q_1 = \{A, B, C, D\}$, $Q_2 = \{B, C, F, D\}$ and $Q_3 = \{F, D, G\}$. Alternatively, if ordering $d_2 = (A, B, C, F, D, G)$ in Figure 5.13c is used, the induced graph generated has only one added edge. The cliques in this case are: $Q_1 = \{G, F\}$, $Q_2 = \{A, B, D\}$, $Q_3 = \{B, C, F\}$ and $Q_4 = \{A, B, C\}$. Yet, another example is given in Figure 5.13d. The corresponding join-trees of orderings d_1 and d_2 are depicted in the earlier decompositions observed in Figure 5.11a and 5.11b, respectively.

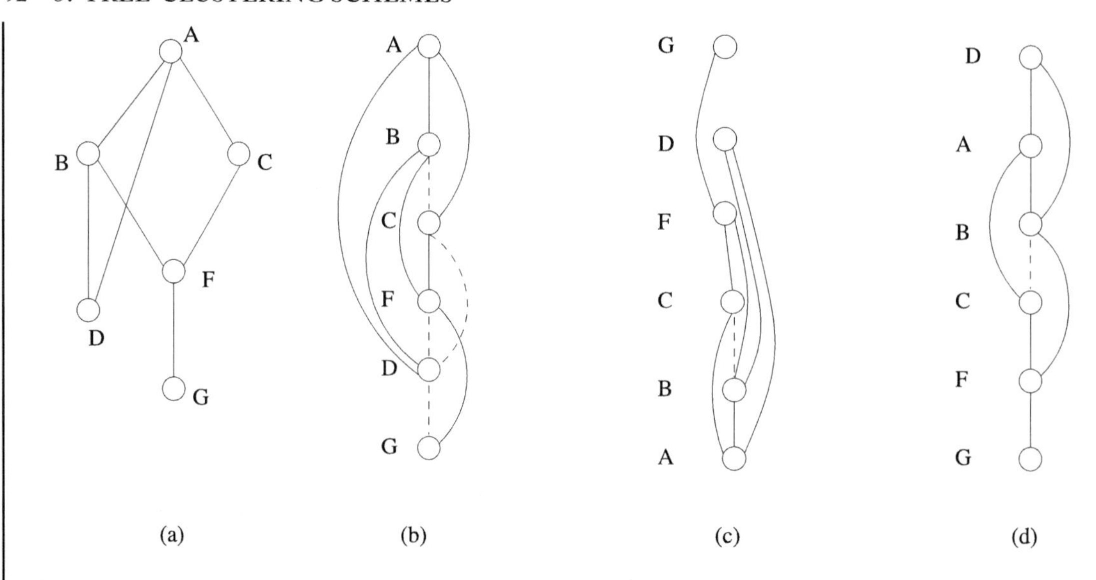

Figure 5.13: A graph (a) and three of its induced graphs (b), (c), and (d).

Treewidth and induced-width. Clearly, if the ordering used by JTC has induced width $w(d)$, the treewidth of the resulting join-tree is $w(d)$ as well. And, vice-versa, given a tree-decomposition of a graph whose treewidth is w, there exist an ordering of the nodes d whose induced-width satisfies $w(d) = w$. In other words *treewidth and induced-width can be viewed as synonym concepts for graphs*, yet induced-width is explicitly defined for an ordering.

5.3 PROPERTIES OF CTE FOR GENERAL MODELS

Algorithm CTE which takes as an input a tree-decomposition of a graphical model, and evidence, outputs an *explicit* tree-decomposition. In this section we will prove this claim and discuss issues of complexity.

Convergence. It is clear that CTE converges after 2 message passing along the tree. If we remove the edge (u, v) we get 2 subtrees. One rooted at node u and one rooted at node v. The outgoing message from u to v depends solely on information in the subtree that include u but not v. Therefore the message $m_{u \to v}$ will not be influenced by the message $m_{v \to v}$.

Theorem 5.23 Convergence of CTE. *Algorithm CTE converges after two iterations.*

5.3.1 CORRECTNESS OF CTE

The correctness of CTE can be shown in two steps. First showing that CTE can solve acyclic graphical models. Since a tree-decomposition transforms a graphical model into an acyclic one, the correctness argument follows. However, instead of taking this route, we will next state the correctness based on the general properties of the *combine* $\bigotimes$ and *marginalize* $\Downarrow$ operators in order to emphasize the broad applicability of this algorithm. The following theorem articulates the properties which are required for correctness. The proof can be found in [Kask and Dechter, 2005] and in the Appendix of this chapter.

Theorem 5.24 Soundness and completeness. *Given a graphical model and reasoning tasks* $\mathcal{M} = \langle \mathbf{X}, \mathbf{D}, \mathbf{F}, \bigotimes, \Downarrow \rangle$, *and assuming that the combination operator* $\bigotimes$ *and the marginalization operator* $\Downarrow_Y$ *satisfy the following properties [Shenoy, 1997]):*

1. *order of marginalization does not matter:*
   $\Downarrow_{X-\{X_i\}} (\Downarrow_{X-\{X_j\}} f(X)) = \Downarrow_{X-\{X_j\}} (\Downarrow_{X-\{X_i\}} f(X))$;
2. *commutativity:* $f \bigotimes g = g \bigotimes f$;
3. *associativity:* $f \bigotimes (g \bigotimes h) = (f \bigotimes g) \bigotimes h$;
4. *restricted distributivity:*
   $\Downarrow_{X-\{X_k\}} [f(X - \{X_k\}) \bigotimes g(X)] = f(X - \{X_k\}) \bigotimes \Downarrow_{X-\{X_k\}} g(X)$.

Algorithm CTE is sound and complete. Namely, it is guaranteed to transform a tree-decomposition $< T, \chi, \psi >$ into an explicit one. Namely, for every node $v \in T$, given the messages generated $m_{u \to v}$ for all $(u, v) \in T$, then

$$M_{\chi(u)} = \psi_u \otimes (\bigotimes_{\{j|(j,u) \in E\}} m_{j \to u}) .$$

where $M_{\chi(u)}$ is explicit relative to $\chi(u)$ (Definition 5.4). For a proof see the Appendix. □

It is common to call the combined function in a cluster as *belief*. Another concept associated with clusters is their normalized constant.

Definition 5.25 belief, normalizing constants. Given a tree decomposition $< T, \chi, \Psi >$, $T = (V, E)$ of a graphical model $\mathcal{M} =< X, D, F, \bigotimes, \Downarrow >$, and a set of messages denoted by m (potentially generated by CTE, but not only) then the beliefs associated with each cluster $u \in T$, relative to incoming messages $\{m\}$ is:

$$b_u = \psi_u \otimes [\bigotimes_{k \in neighbors(v)} m_{k \to u}] \tag{5.6}$$

We also define

$$b_{sep(u,v)} = \Downarrow_{\chi(u)-sep(u,v)} b_u. \tag{5.7}$$

The normalizing constant of a node v is defined by:

$$K_u = \Downarrow_{\chi(u)} b_u \tag{5.8}$$

What we showed is that if the set of messages m were generated by CTE then the beliefs of each node is the explicit function of the model. Clearly, therefore for any two adjacent nodes in the tree decomposition, marginalizing the beliefs over the separators must yields identical function which is the explicit function over that separator. In the case of probabilities this means that the marginal probability on the separator variables can be obtained in either one of the clusters.

Definition 5.26 Pairwise consistency. Given a tree decomposition $< T, \chi, \psi >$, $T = (V, E)$ of $\mathcal{M} = \langle \mathbf{X}, \mathbf{D}, \mathbf{F}, \bigotimes, \Downarrow \rangle$, and a set of messages $m_{u \to v}, m_{v \to u}$, for every edge $(u, v) \in T$, then

$$\Downarrow_{sep(u,v)} [\psi_v \otimes (\bigotimes_{\{j|(j,v) \in E\}} m_{j \to v})] = \Downarrow_{sep(u,v)} [\psi_u \otimes (\bigotimes_{\{j|(j,u) \in E\}} m_{j \to u})].$$

using the definition of beliefs this is equivalent to:

$$\Downarrow_{sep(u,v)} b_v = b_{uv} = \Downarrow_{sep(u,v)} b_u.$$

Interestingly, we can prove pairwise consistency of CTE without using explicitness, as well as several properties and in particular that the normalizing constants of all nodes are identical.

Theorem 5.27 *Given a tree decomposition $< T, \chi, \psi >$, $T = (V, E)$ of $\mathcal{M} = \langle \mathbf{X}, \mathbf{D}, \mathbf{F}, \bigotimes, \Downarrow \rangle$, when CTE terminates with the set of messages $m_{u \to v}, m_{v \to u}$, then for any edge in T,*

1. *The message obey symmetry, namley:*

$$b_{sep(u,v)} = m_{u \to v} \otimes m_{v \to u} \tag{5.9}$$

2. *The belief generated are pairwise consistent*
3. *The normalizing constant is unique. Namely,*

$$K_u = K_v \tag{5.10}$$

Proof. We will prove 1, and from 1 we prove 2 and from 2 we prove 3. For readability we use product $\prod$ for the combination operator $\bigotimes$ and summation $\sum$ for the marginalization $\Downarrow$.
1. We have by definition that

$$m_{u \to v} = \sum_{\chi(u)-sep(u,v)} \psi_u \prod_{r \in ne(u), r \neq v} m_{r \to u}$$

Multiplying both sides by $m_{v \to u}$ we get

$$m_{u \to v}(x_{uv}) \cdot m_{v \to u}(x_{vu}) = \sum_{\chi(u)-sep(u,v)} \psi(x_u) \prod_{r \in ne(u)} m_{r \to u}(x_{ru}) = = \sum_{\chi_u - sep(u,v)} b_u = b_{sep(u,v)} \tag{5.11}$$

which yields symmetry.
2. The property pwc follows immediately from symmetry (show as an exercise.)
3. (proof of normal constant.) If $< T, \chi, \Psi >$ is pwc relative to messages generated by CTE then

$$K_u = \sum_{\chi(u)} b_u = \sum_{sep(u,v)} \sum_{\chi(u)-\chi(uv)} b_u =$$

and because of pairwise consistency

$$= \sum_{sep(u,v)} b_{uv} = \sum_{sep(u,v)} b_{vu} = \sum_{sep(u,v)} \sum_{\chi(v)-sep(u,v)} b_v = \sum_{\chi(v)} b_v = K_v$$

□

If our graphical model and query are of a sum-product type, the normalizing constant is the probability of evidence or the partition function. And, as expected, we can compute this in any node in the tree decomposition. If it is the max-product or min-sum model, the normalizing constant is the cost of an optimal solution, and it also can be derived in any dode.

5.3.2 COMPLEXITY OF CTE

Algorithm CTE can be subtly varied to influence its time and space complexities. The description in Figure 5.10 seems to imply an implementation whose time and space complexities are the same. Namely, that the space complexity must also be exponential in the induced-width or treewidth, denoted w. Indeed, if we compute the message in the equation in Fig. 5.10 in a brute-force manner, recording the *combined function* first, and subsequently marginalizing over the separator, we will have space complexity exponential in w.

However, we can, instead, interleave the combination and marginalization operations, and thereby make the space complexity identical to the size of the sent message only, as follows.

In words, for each assignment $\mathbf{x}$ to the variables in $\chi(u)$, we compute the product functional value, and accumulate the sum value on the separator, sep, updating the message function

GENERATE MESSAGES

Input: cluster u, its $\chi(u)$ and $\psi(u)$, its neighbor v with $\chi(v)$, $sep = \chi(u) \cap \chi(v)$.
1. initialize: for all $\mathbf{x}_{sep}$, $m_{u \to v}(\mathbf{x}_{sep}) \leftarrow 0$.
2. **for** every assignment $\mathbf{x}_{\chi(u)}$, **do**
3. $m_{u \to v}(\mathbf{x}_{sep}) \Leftarrow m_{u \to v}(\mathbf{x}_{sep}) + \psi_u(\mathbf{x}_{\chi(u)}) \cdot \prod_{\{j|(j,u) \in T, j \neq v\}} m_{j \to u}(\mathbf{x}_{sep})$
4. **end for**

Figure 5.14: Generate messages.

$m_{u \to v}(\mathbf{x}_{sep})$. With this modification we now can state (and then prove) the general complexity of CTE (for a proof see Appendix).

Theorem 5.28 Complexity of CTE. *Given a graphical model $\mathcal{M} = \langle \mathbf{X}, \mathbf{D}, \mathbf{F}, \bigotimes, \Downarrow \rangle$ and its* tree-decomposition $< T, \chi, \psi >$, *where $T = (V, E)$, if N is the number of vertices in V, w its tree-width, sep its maximum separator, r the number of functions in F, deg the maximum degree in T, and k the maximum domain size of a variable, the time complexity of CTE is $O((r + N) \cdot deg \cdot k^{w+1})$ and its space complexity is $O(N \cdot k^{|sep|})$ (for a proof see appendix.)*

Trading space for time in CTE. As we noted earlier, given any tree decomposition we can generate new tree decompositions by merging adjacent clusters. While time complexity will increase, this process can generate smaller separators, and therefore smaller memory.

Example 5.29 Consider the tree decompositions in Figure 5.9. For the first two decompositions CTE will have time exponential in 3 and space complexity exponential in 2. The third yields time exponential in 5 but space exponential in 1.

5.4 ILLUSTRATION OF CTE FOR SPECIFIC MODELS

In this last section of this chapter we will provide more details on algorithms tailored to specific graphical models such as Bayesian networks and constraint networks.

5.4.1 BELIEF UPDATING AND PROBABILITY OF EVIDENCE

As we saw, applying algorithm CTE to Bayesian networks when combination is *product* and the marginalization operators is *summation*, yields an algorithm that computes the explicit clusters for a given tree decomposition. In this case the explicit functions are the posterior marginal probability distribution given the evidence over the cluster's variables. Therefore, when the algorithm

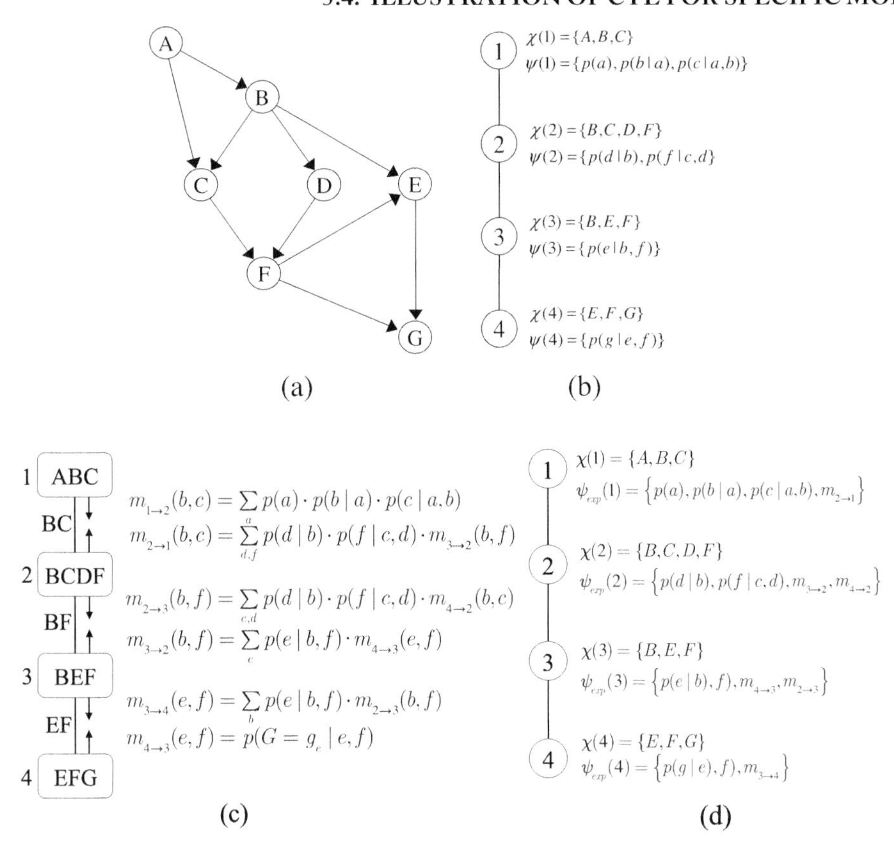

Figure 5.15: [Execution of CTE-bel]: (a) a belief network; (b) A join-tree decomposition; (c) execution of CTE-bel; and (d) the explicit tree-decomposition.

terminates the marginals can be obtained by the normalized product over all functions in the corresponding clusters. From these clusters one can also compute the probability of evidence or the posterior beliefs over singleton variables. We will refer to this specialized algorithm as *CTE-bel*. When a cluster sends a message to a neighbor, the message may contains a single *combined* function and may also contain *individual* functions that do not share variables with the relevant eliminator.

Example 5.30 Figure 5.15 describes a belief network (a) and a join-tree decomposition for it (b). Figure 5.15c shows the trace of running *CTE-bel*. In this case no individual functions appear between any of the clusters. Figure 5.15d shows the explicit output tree-decomposition. If we want to compute the probability of evidence $P(G = g_e)$, we can pick cluster 4, for example, and compute $P(G = g_e) = \sum_{e,f,g=g_e} P(g|e, f) \cdot m_{3\to4}(e, f)$ and if we wish to compute the belief for variable B for example we can use the second or the first bucket, $P(B|g_e) =$

$\alpha \cdot \sum_{a,c} P(a) \cdot p(b|a) \cdot p(c|a,b) \cdot m_{2\to1}(b,c)$ where α is the normalizing constant that is 1 divided by the probability of evidence.

Pearl's Belief Propagation over Polytrees

A special acyclic graphical models are *polytrees*. This case deserves attention for historical reasons; it was recognized by Pearl [Pearl, 1988] as a generalization of trees on which his known belief propagation algorithm was presented. It also gives rise to an iterative approximation algorithm over general networks, known as *Iterative BP* or *loopy BP* [Weiss and Pearl, 2010].

Definition 5.31 Polytree. A polytree is a directed acyclic graph whose underlying undirected graph has no cycles (see Figure 5.16a).

It is easy to see that the dual graph of a polytree is a tree, and thus yields an acyclic problem that has a join-tree decomposition where each family resides in a single node u in the decomposition. Namely $\chi(u) = \{X\} \cup pa(X)$, and $\psi(u) = \{P(X|pa(X))\}$. Note, that the separators in this *polytree decomposition*, are all singleton variables. In summary,

Proposition 5.32 *A polytree has a dual graph which is a tree and it is therefore an acyclic graphical model (Prove as an exercise.)*

It can be shown that Pearl's BP is identical to CTE if applied to the poly-tree based dual tree that is rooted in accordance with the poly-tree's topological order and where in one direction the CTE messages are named λs and in the reverse direction they are named π's.

Example 5.33 Consider the polytree given in Figure 5.16a. An ordering along which we can run CTE is given in Figure 5.16b, a directed polytree decomposition is given in Figure 5.16c along with the π and λ messages. The explicit output tree-decomposition is given in part 5.16d. Once the propagation terminates, beliefs can be computed in each cluster.

5.4.2 CONSTRAINT NETWORKS

Algorithm CTE for constraint networks can be obtained straightforwardly by using the join operation for combination and the relational project for marginalization. The explicit algorithm called *CTE-cons* is given in Figure 5.17 (see also 9.10 in [Dechter, 2003]). It yields an *explicit* representation of the constraints in each node. This makes it possible to answer most relevant queries locally, by consulting the constraints inside each of these nodes only. This property of explicitness is called *minimality* and *decomposability* in [Montanari, 1974]as defined next.

Definition 5.34 Minimal subproblem, a decomposable network. Given a constraint problem $\mathcal{R} = (X, D, C)$, where $C = \{R_{S_1}, ..., R_{S_m}\}$ and a subset of variables $Y \subseteq X$, a subproblem over

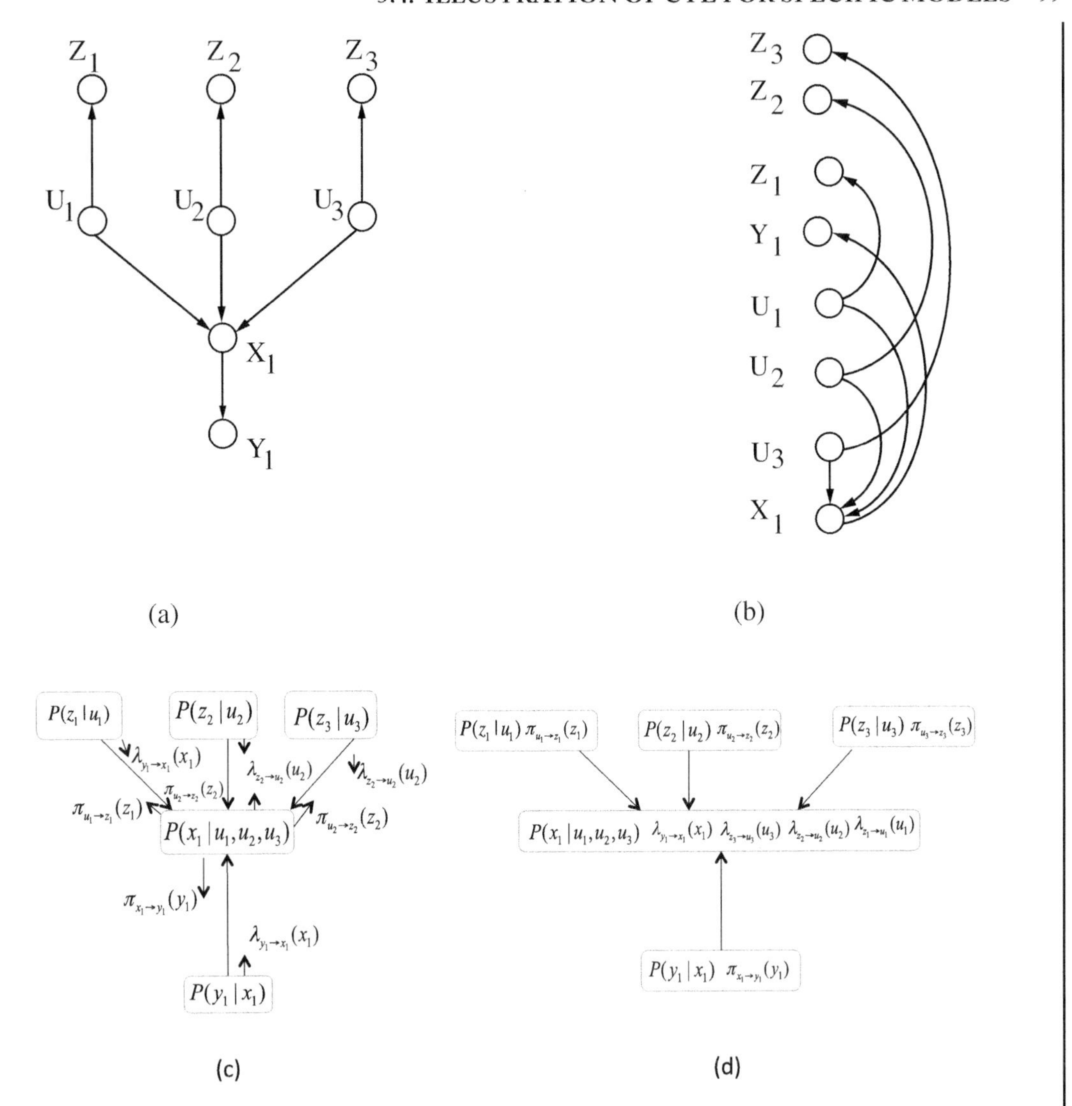

Figure 5.16: (a) A polytree and (b) a legal processing ordering, (c) a polytree decomposition and messages, and (d) the explicit decomposition.

$Y, \mathcal{R}_Y = (Y, D_Y, C_Y)$ is minimal relative to $\mathcal{R}$, iff $sol(R_{\mathbf{Y}}) = \pi_Y sol(\mathcal{R})$ where $sol(\mathcal{R}) = \bowtie_{\mathbf{R}\in\mathcal{R}} \mathbf{R}$ is the set of all solutions of network $\mathcal{R}$. A network of constraints is decomposable if each of its subnetworks is minimal.

Cluster tree-elimination (CTE-cons)
Input: A tree decomposition $< T, \chi, \psi >$ for a constraint network $\mathcal{R} =< X, D, C, \bowtie, \pi >$.
ψ_u is the join of original relations in cluster u
Output: An explicit tree-decomposition where each cluster is minimal (i.e., explicit).
Compute messages:
for every edge (u, v) in the tree, do

- Let $m_{u\rightarrow v}$ denote the message sent from u to v.

$$m_{u\rightarrow v} \leftarrow \pi_{sep(u,v)}(\psi_u \bowtie (\bowtie_{R_i \in cluster_v(u)}) R_i) \tag{5.12}$$

endfor
Return: A tree-decomposition augmented with constraint messages. For every node $u \in T$, return the decomposable (explicit) subproblem $\psi^{'}(u) = \psi(u) \cup \{m_{(i\rightarrow u)} | (i, u) \in T\}$.
If $\psi^{'}$ is consistent for every node the problem is consistent.
Generate a solution in a backtrack-free manner.

Figure 5.17: Algorithm cluster-tree elimination (CTE).

An immediate illustration can be generated from the example in Figure 5.11. All we need is to assume that the functions are relations and the product is replaced by a *join* $\bowtie$ operator while the sum operator is replaced with the relational projection π. >From the comrrectness of CTE it follows that:

Theorem 5.35 *Given a tree decomposition* $< T, \chi, \psi >$ *for a constraint network* $\mathcal{R} = \langle \mathbf{X}, \mathbf{D}, \mathbf{C} = \{R_{S_1}, ..., R_{S_r}\}, \bowtie\rangle$ *where* R_{S_i} *is the relation over a scope* S_i. *Then at termination of* CTE-cons, *the constraints in each node constitute a minimal network. Namely, for each node* u, $\bowtie_{R_i \in cluster(u)} R_i = \pi_{\chi(u)}(\bowtie_{R_i \in \mathcal{R}} R_i)$.

The minimality property is powerful. Once we have a minimal subnetwork which is not empty we immediately know that the network is consistent. Moreover, the resulting tree-decomposition whose clusters are minimal can be shown to be backtrack-free along orderings consistent with the tree structure. This means that we can generate a solution in any order along the tree and we are guaranteed to not have any dead-ends. Therefore solution generation is linear in the output network. (Exercise: prove that *CTE-cons* generates a backtrack-free ordering along some variable orderings.) Interestingly, in the case of constraints we do not need to be careful regarding the exclusion of the message sent from u to v when v computes the message it sends to u.

Counting. We have looked at algorithms for solving the tasks of computing the probability of evidence or marginals in probabilistic networks and of answering constraint satisfaction tasks over constraint networks. The first is often referred to as the *sum-product* algorithm and the second as the *join-project*. The join-project algorithm can be accomplished using Boolean operators. If we express the relations as $\{0, 1\}$ cost functions("0" for inconsistent tuple and "1" for a consistent tuple), combining functions by a Boolean product operators and marginalizing using Boolean summation will yield an identical algorithm to *CTE-cons*. If we use regular summation for the marginalization operator and a regular product operator for combination over this (0,1) cost representation of relation, then the *CTE* sum-product algorithm computes the *number of solutions* of the constraint network. Moreover, it associates any partial configuration in each cluster with the number of solutions extending it. We will refer to the algorithm as *CTE-count*.

Definition 5.36 Counts of partial solutions. Given a $\mathcal{R} =< X, D, C >$, and given a partial value assignment $\mathbf{x_S}$ over scope $\mathbf{S}$, $\mathbf{S} \subseteq \mathbf{X}$, we denote by $count(\mathbf{x_S})$, the number of full solutions of $\mathcal{R}$ that extend the partial assignment $\mathbf{x_S}$. Namely,

$$count(\mathbf{x_S}) = |\{\mathbf{y} \in sol(\mathcal{R}) | \mathbf{y_S} = \mathbf{x_S}\}| \, .$$

Theorem 5.37 *Given a tree decomposition $< T, \chi, \psi >$ for a constraint network $\mathcal{R} =< X, D, C = \{C_1, ..., C_r\} >$, where each C_i is represented as a zero-one cost function. Then when* CTE-count *terminates with $< T, \chi, \psi' >$, then for each u and for each of its assignment $\mathbf{x}_{\chi(u)}$ over its scope, $\prod_{f \in \psi'(u)} f = count(\mathbf{x}_{\chi(u)})$.*

5.4.3 OPTIMIZATION

A popular type of CTE algorithm is for solving an optimization query. Namely when $\Downarrow$ is max or min. If, for example, we want to solve the MPE task over Bayesian networks, we know that we can do so by the bucket elimination algorithm when in each bucket we use the max-product combination operators. Extending that into a *CTE-max* (or a *CTE-min*, when we combine functions by summation) algorithm (called often the max-product algorithm) will generate, not only the maximum cost, but also for every partial assignment of a node u, $\mathbf{x}_{\chi(u)}$, in a tree decomposition, the maximum cost (e.g., probability) of any of its completions into a full assignment. Finally, if in our cost network the combination operator is summation, we can use min-sum or max-sum CTE. In all these cases, at termination we have what is sometimes called, *the optimal cost to go* associated with *each node* u, and therefore with each partial assignment $\mathbf{x}_{\chi(u)}$ in the tree decomposition.

For marginal map queries the extension is straightforward, we only need to make sure the the tree-decomposition will be along ordering that are restricted. Once the clusters are generated, as join-trees for example, the messages are generated in the same manner when summation variables in each cluster should be eliminated before maximization variables.

5.5 SUMMARY AND BIBLIOGRAPHICAL NOTES

Join-tree clustering was introduced in constraint processing by Dechter and Pearl [Dechter and Pearl, 1989] and in probabilistic networks by Spigelhalter et al. [Lauritzen and Spiegelhalter, 1988]. Both methods are based on the characterization by relational-database researchers that acyclic-databases have an underlying tree-structure, called join-tree, that allows polynomial query processing using join-project operations and easy identification procedures [Beeri *et al.*, 1983; Maier, 1983; Tarjan and Yannakakis, 1984]. In both constraint networks and belief networks, it was observed that the complexity of compiling any knowledge-base into an acyclic one is exponential in the cluster size, which is characterized by the induced width or tree width. At the same time, variable-elimination algorithms developed in [Bertele and Brioschi, 1972; Seidel, 1981] and [Dechter and Pearl, 1987] (e.g., adaptive-consistency and bucket-elimination) were also observed to be governed by the same complexity graph-parameter. In [Dechter and Pearl, 1987, 1989] the connection between induced-width and treewidth was recognized via the work of [Arnborg, 1985] on treewidth, k-trees and partial k-trees. This was made explicit later in [Freuder, 1992]. The similarity between variable-elimination and tree-clustering from the constraint perspective was analyzed by [Dechter and Pearl, 1989]. Independently of this investigation, the treewidth parameter was undergoing intensive investigation in the theoretic-graph-community. It characterizes the best embedding of a graph or a hypergraph in a hypertree. Various connections between hypertrees, chordal graphs and k-trees were made by Arnborg et al. [Arnborg, 1985; S. A. Arnborg and Proskourowski, 1987]. They showed that finding the smallest treewidth of a graph is NP-complete, but deciding if the graph has a treewidth below a certain constant k is polynomial in k. A recent analysis shows that this task can be accomplished in $O(n \cdot f(k))$ where $f(k)$ is a very bad exponential function of k. [Bodlaender, 1997]. The style of describing a tree-decomposition is adopted from [Georg Gottlob and Scarcello, 2000] where they talk about hypertree decompositions (not used here).

5.6 APPENDIX: PROOFS

Proof of Theorem 5.14

The algorithm is linear because there exists an ordering for which each function resides alone in its bucket, and for which BTE generates messages that are subsumed by the original functions' scopes. Clearly, such messages (at most n in each direction) can be generated in time and space bounded by the functions' sizes (i.e., number of tuples in the domain of each input function). A desired ordering can be generated by processing leaf nodes along the join-tree of the acyclic model, imposing a partial. We can show (exercise) that the ordering generated facilitates messages whose scopes are subsumed by the original function scopes. This implies a linear complexity. □

Proof of Theorem 5.19

Given bucket tree $T = (V, E)$ of $\mathcal{M}$, whose nodes are mapped to buckets, we need to show how the tree can be associated with mappings χ and ψ that satisfy the that conditions of Definition 5.15. In other words, the tree structure T in tree decomposition $< T, \chi, \psi >$ is the bucket tree structure, where each B_i corresponds to a vertex in V. If a bucket B_i has a parent (i.e., is connected to) bucket B_j, there is an edge $(B_i, B_j) \in E$. Labeling $\chi(B_i)$ is defined to be the union of the scopes of new and old functions in B_i during processing by BTE, and labeling $\psi(B_i)$ is defined to be the set of functions in the initial partition in B_i. With these definitions, condition 1 of Definition 5.15 is satisfied because each function is placed into exactly one bucket and condition 2 of Definition 5.15 is also satisfied because labeling $\chi(B_i)$ is the union of scopes of all functions in B_i. Condition 4 of Definition 5.15 is trivially satisfied since there is exactly one bucket for each variable.

Finally, we need to prove the connectedness property. Let's assume that there is a variable X_k with respect to which the connectedness property is violated. This means that there must be (at least) two disjoint subtrees, T_1 and T_2, of T, such that each vertex in both subtrees contains X_k, and there is no edge between a vertex in T_1 and T_2. Let B_I be a vertex in T_1 such that X_i is the earliest relative to ordering d, and B_j a vertex in T_2 such that X_j is the earliest in ordering d. Since T_1 and T_2 are disjoint, it must be that $X_i \neq X_j$. However, this is impossible since this would mean that there are two buckets that eliminate variable X_k. □

Proof of Theorem 5.28

The time complexity of processing a vertex u in tree T is $deg_u \cdot (|\psi(u)| + deg_u - 1) \cdot k^{|\chi(u)|}$, where deg_u is the degree of u, because vertex u has to send out deg_u messages, each being a combination of $(|\psi(u)| + deg_u - 1)$ functions, and requiring the enumeration of $k^{|\chi(u)|}$ combinations of values. The time complexity of CTE is

$$Time(CTE) = \sum_u deg_u \cdot (|\psi(u)| + deg_u - 1) \cdot k^{|\chi(u)|} .$$

By bounding the first occurrence of deg_u by deg and $|\chi(u)|$ by $w + 1$, we get

$$Time(CTE) \leq deg \cdot k^{w+1} \cdot \sum_u (|\psi(u)| + deg_u - 1) .$$

Since $\sum_u |\psi(u)| = r$ we can write

$$\begin{aligned} Time(CTE) &\leq deg \cdot k^{w+1} \cdot (r + N) \\ &= O((r + N) \cdot deg \cdot k^{w+1}) . \end{aligned}$$

For each edge CTE will record two functions. Since the number of edges is bounded by N and the size of each function we record is bounded by $k^{|sep|}$.

If the cluster tree is minimal (for any u and v, $sep(u,v) \subset \chi(u)$ and $sep(u,v) \subset \chi(v)$), then we can bound the number of vertices N by n. Assuming $r \geq n$, the time complexity of CTE applied to a minimal tree-decomposition is $O(deg \cdot r \cdot k^{w+1})$. □

Proof of Theorem 5.24

Using the four properties of combination and marginalization operators, the claim can be proved by induction on the depth of the tree. Specifically, we will show that for every node u, CTE generates the explicit representation for that node. Namely, for every node u, $\bigotimes_{f \in \bar{\psi}(u)} f = \mathcal{M}_{\chi(u)}$.

Let $< T, \chi, \psi >$, where $T = (V, E)$ be a cluster-tree decomposition for $\mathcal{M}$, and let root it in vertex $v \in V$. We can create a partial order of the vertices of T along the rooted tree. We denote by $T_u = (V_u, E_u)$ the subtree rooted at vertex u and define by $\chi(T_u)$ all the variables associated with nodes appearing in T_u, namely: $\chi(T_u) = \bigcup_{v \in V_u} \chi(v)$.

Since commutativity permits combining the functions in any order, we select an ordering of the nodes in T $d(j) \in V, j = 1, ..., |V|$ where a vertex in the rooted tree T precedes its children in the ordering, and thus the first vertex is the root of the tree is v. As usual we denote by $\psi_u = \bigotimes_{f \in \psi(u)} f$, the combination of all the input functions in node u. Because of associativity and commutativity, clearly:

$$\forall\, u \;\; \mathcal{M}_{\chi(u)} = \Downarrow_{\chi(u)} \bigotimes_{j=1}^{|V|} \psi_{d(j)}.$$

Let u be a node having the parent w in the rooted tree T, and define $elim(u) = \chi(u) - sep(u,w)$ and $elim(T_u) = \bigcup_{v \in V_u} elim(v)$. (We will show that $elim(T_u)$ is the set of variables that are eliminated by CTE in the subtree rooted at u when sending a message to parent w). Because of the connectedness property, variables in $elim(T_u)$, appear only in the subtree rooted at u. In other words, $elim(T_u) \bigcap \{X_i | X_i \in V - \chi(T_u)\} = \emptyset$. Consequently, we can marginalize ($\Downarrow$) over such variables earlier in the process of deriving $\mathcal{M}_{\chi(u)}$ (note that $\Downarrow_{Z_i}$ means marginalizing over $X - Z_i$). If $X_i \notin \chi(u)$ and if $X_i \in elim(d(k))$ for some k, then, the marginalization eliminating X_i can be applied to $\bigotimes_{j=k}^{|V|} \psi_{d(j)}$ instead of to $\bigotimes_{j=1}^{|V|} \psi_{d(j)}$. This is safe to do, because as shown above, if a variable X_i belongs to $elim(d(k))$, then it cannot be part of any $\psi_{d(j)}$, $j < k$. We can therefore derive $\mathcal{M}_{\chi(u)}$ as follows:

$$\mathcal{M}_{\chi(u)} = \Downarrow_{\chi(u)} \bigotimes_{j=1}^{|V|} \psi_{d(j)} = \tag{5.13}$$

(because of the tree structure)

$$= \Downarrow_{\chi(u)} [\bigotimes_{j=1}^{d(k-1)} \psi_{d(j)} \Downarrow_{(X - elim(d(k)))} \bigotimes_{j=k}^{|V|} \psi_{d(j)}] = \tag{5.14}$$

$$= \Downarrow_{\chi(u)} \bigotimes_{j=1}^{d(k-1)} \psi_{d(j)} \otimes F_T(d(k))\,, \tag{5.15}$$

where for $u = d(k)$ $F_T(u) = \Downarrow_{(X-elim(u))} \bigotimes_{j=k}^{|V|} \psi_{d(j)}$. (note that $(X - elim(u) = sep(u, w))$. We now assert that due to properties 1-4, $F_T(u)$ obeys a recursive definition relative to the tree-decomposition and that this recursive definition is identical to the messages computed by CTE and sent from u to its parent w. This is articulated by the following proposition and will conclude the proof.

Proposition 5.38 *The functions $F_T(u)$ defined above relative to the rooted tree-decomposition T, obey the following recursive definition. Let $ch(u)$ be the set of children of u in the rooted tree T.*

- *If $ch(u) = \emptyset$ (vertex u is a leaf vertex), then $F_T(u) = \Downarrow_{(X-elim(u))} \psi_u$.*
- *Otherwise, $F_T(u) = \Downarrow_{(X-elim(u))} \psi_u \otimes \bigotimes_{w \in ch(u)} F_T(w)$.*

It is easy to see that the messages computed by CTE up the tree decomposition along T are the $F_T(u)$ functions. Namely, For every node u and its parent w, $F_T(u) = m_{u \to w}$, and in particular at the root node v $F_T(v) = \mathcal{M}_{\chi(v)}$ which is identical to the message v can send to its (empty parent).

This completes the proof for the root node v. Since the argument can be applied to any node that can be made into a root of the tree, we have explicitness for all the nodes in the tree-decomposition. □

CHAPTER 6

AND/OR Search Spaces and Algorithms for Graphical Models

In this chapter we start the discussion of the second type of reasoning algorithms, those that are based on the *conditioning* step, namely, on assigning a single value to a variable. To recall, algorithms for processing graphical models fall into two general types: inference-based and search-based. Inference-based algorithms (*e.g.*, variable-elimination, tree-clustering discussed earlier) are good at exploiting the independencies displayed by the underlying graphical model and in avoiding redundant computation. They have worst case time guarantee which is exponential in the treewidth of the graph. Unfortunately, any method that is time-exponential in the treewidth is also space exponential in the treewidth or in the related separator-width parameter and therefore, not feasible for models that have large treewidths.

Traditional search algorithms (*e.g.*, depth-first branch-and-bound, best-first search) traverse the model's search space where each path represents a partial or a full solution. For example, we can compute expression 6.1 for $P(G = 0, D = 1)$ of the network of Figure 5.8(a) by traversing the search-tree in Figure 6.1 along an ordering, from first variable to last variable.

$$P(D = 1, G = 0)) = \sum_{a,c,b,f,d=1,g=0} P(g = 0|f)P(f|b,c)P(d = 1|a,b)P(c|a)P(b|a)P(a) \tag{6.1}$$

$$= \sum_a P(a) \sum_c P(c|a) \sum_b P(b|a) \sum_f P(f|b,c)P(d = 1|b,a)P(g = 0|f).$$

Specifically, the arcs of each path are weighted by numerical values extracted from the CPT's of the problem that correspond to the variable assignments along the path. The bottom path shows explicitly the functions at each arc and the leaves provide the probabilities conditioned on the evidence as is shown. In this traditional search tree, every complete path expresses a *solution*, namely a full assignment to the variables and the product of its weight gives the probability of this *solution*.

The search tree can be traversed by depth-first search accumulating the appropriate sums of probabilities (details will be given shortly). It can also be *searched* to find the assignment having the highest probability, thus solving the *mpe* task.

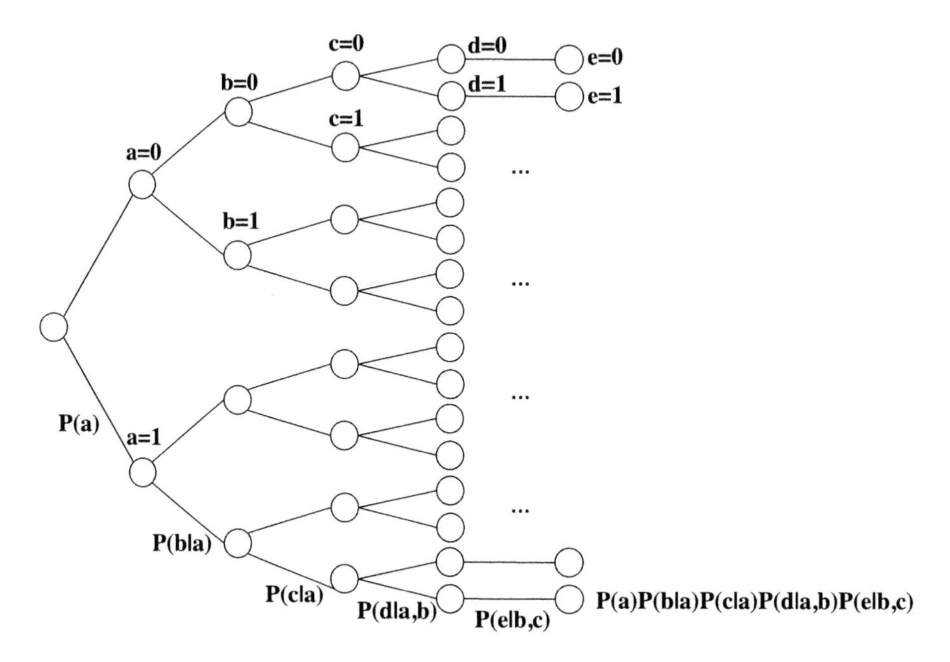

Figure 6.1: Probability tree for computing P(d=1,g=0).

Notice that the structure of search spaces does not retain the independencies represented in the underlying graphical model and may lead to inferior schemes compared with inference algorithms. The size of this search tree is $O(k^n)$ when k bounds the somain size and n is the number of variables. On the other hand, the memory requirements of search algorithms may be less severe than those of inference algorithms; if we use *DFS* traversal it can be accomplished with linear memory. Furthermore, search requires only an implicit, generative, specification of the functions (given in a procedural or functional form) while inference schemes often rely on an explicit tabular representation over the (discrete) variables. For these reasons search algorithms are the only choice available for models with large treewidth, large domains, and with implicit representation.

In this chapter we will show that it is beneficial to depart from the standard linear search space in favor of AND/OR search spaces, originally introduced in the context of heuristic search [Nillson, 1980], primarily because they encode some of the structural information in the graphical models. In particular, AND/OR search spaces can capture the independencies in the graphical model to yield AND/OR search trees that are exponentially smaller than the standard search tree, which we call *OR* tree. We will provide analysis of the size of the AND/OR search tree and show that it is bounded exponentially by the height of some tree that spans the graphical model. Subsequently, we show that the search tree may contain significant redundancy that when identified, can be removed yielding AND/OR search graphs. This additional savings can reduce the size of the AND/OR search space further to the point that it can be guaranteed to be no larger than exponentially in the graphical model treewidth,

6.1 AND/OR SEARCH TREES

We will present and contrast the concepts of *OR* vs. *AND/OR* search spaces of *graphical models* starting with an example of a constraint network.

Example 6.1 Consider the simple tree graphical model (*i.e.*, whose primal graph is a tree) in Figure 6.2a, over domains of variables $\{1, 2, 3\}$, which represents a graph-coloring problem. Namely, each node should be assigned a value such that adjacent nodes have different values. The common way to solve this problem is to consider all partial and full solutions to the variables by traversing the problem's search tree, in which each partial path is an assignment to a subset of the variables, and the solutions are paths of length n when n is the number of variables. The problem depicted in Figure 6.2a yields the OR search tree in Figure 6.2b.

Notice, however, that once variable X is assigned the value 1, the search space it roots can be decomposed into two independent subproblems, one that is rooted at Y and one that is rooted at Z, both of which can to be solved independently. Indeed, given $X = 1$, the two search subspaces do not interact. The same decomposition can be associated with the other assignments to X, $(X = 2)$ and $(X = 3)$. Applying the decomposition along the tree (in Figure 6.2a yields the AND/OR search tree in Figure 6.2c. The AND nodes denote problem-decomposition. They indicate that child nodes of an AND nodes can be solved independently. Indeed, in the AND/OR space, a full assignment to all the variables is not a path but a subtree. Comparing the size of the traditional *OR* search tree in Figure 6.2b against the size of the AND/OR search tree, the latter is clearly smaller. The OR search space has $3 \cdot 2^7$ nodes while the AND/OR one has $3 \cdot 2^5$.

More generally, if k is the domain size, a balanced binary tree graphical model (e.g., a graph coloring problem) with n nodes has an OR search tree of size $O(k^n)$. The AND/OR search tree, whose underlying tree graphical model has depth $O(\log_2 n)$, has size $O((2k)^{\log_2 n}) = O(n \cdot k^{\log_2 n}) = O(n^{1+\log_2 k})$. When $k = 2$, this becomes $O(n^2)$ instead of 2^k.

The AND/OR space is not restricted to tree graphical models as in the above example. As we show, it only has to be guided by a tree spanning the primal graph of the model that obeys some conditions to be defined in the next subsection. We define the AND/OR search space relative to a guiding spanning tree of the primal graph.

Definition 6.2 AND/OR search tree. Given a graphical model $\mathcal{M} = \langle \mathbf{X}, \mathbf{D}, \mathbf{F}, \bigotimes \rangle$, its primal graph G and a guiding spanning tree $\mathcal{T}$ of G, the associated AND/OR search tree, denoted

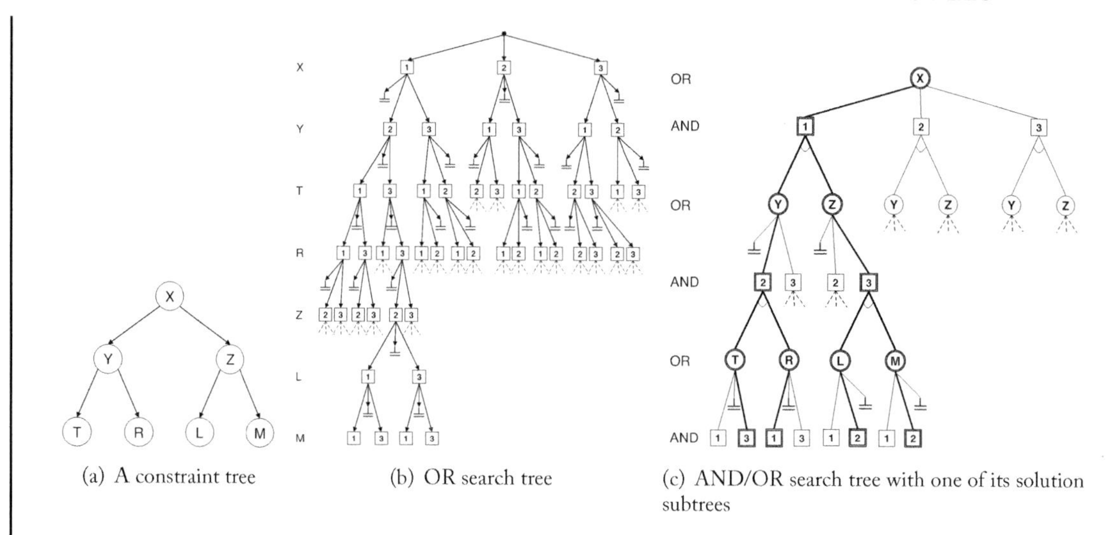

(a) A constraint tree

(b) OR search tree

(c) AND/OR search tree with one of its solution subtrees

Figure 6.2: OR vs. AND/OR search trees; note the connector for AND arcs.

$S_{\mathcal{T}}(\mathcal{M})$, has alternating levels of AND and OR nodes. The OR nodes are labeled X_i and correspond to the variables. The AND nodes are labeled $\langle X_i, x_i \rangle$ (or simply x_i) and correspond to the value assignments of the variables. The structure of the AND/OR search tree is based on the underlying spanning tree $\mathcal{T}$. Its root is an OR node labeled by the root of $\mathcal{T}$.

A path from the root of the $S_{\mathcal{T}}(\mathcal{M})$ to a node n is denoted by $path(n)$. If n is labeled X_i or x_i the path will be denoted $path(n = X_i)$ or $path(n = x_i)$, respectively. The assignment sequence along $path(n)$, denoted $val(path(n))$ is the tuple of values assigned to the variables along the path. That is, the sequence of AND nodes along $path(n)$:

$$\begin{aligned} val(path(n = X_i)) &= \{\langle X_1, x_1 \rangle, \langle X_2, x_2 \rangle, \ldots, \langle X_{i-1}, x_{i-1} \rangle\} = \mathbf{x}_{(1..i-1)}, \\ val(path(n = x_i)) &= \{\langle X_1, x_1 \rangle, \langle X_2, x_2 \rangle, \ldots, \langle X_i, x_i \rangle\} = \mathbf{x}_{(1..i)}. \end{aligned}$$

The set of variables associated with OR nodes along path $path(n)$ is denoted by $var(path(n))$: $var(path(n = X_i)) = \{X_1, \ldots, X_{i-1}\}$, $var(path(n = x_i)) = \{X_1, \ldots, X_i\}$. The parent-child relationship between nodes in the search space are defined as follows.

1. An OR node, n, labeled by X_i has a child AND node, m, labeled $\langle X_i, x_i \rangle$ iff $\langle X_i, x_i \rangle$ is consistent with the assignment $val(path(n))$. Consistency is defined relative to the constraints when we have a constraint problem, or relative to the flat constraints extracted from the zeros in the CPT tables otherwise.

2. An AND node m, labeled $\langle X_i, x_i \rangle$ has a child OR node r labeled Y, iff Y is a child of X in the guiding spanning tree $\mathcal{T}$. Each OR arc emanating from an OR to an AND node is associated with a weight to be defined shortly (see Definition 6.8).

A solution in an AND/OR space is a subtree rather than a path.

Definition 6.3 Solution subtree. A *solution subtree* of an AND/OR search tree contains the root node. For every OR node, if it is in the solution tree then the solution contains one of its child nodes and for each of its included AND nodes the solution contains all its child nodes, and all its leaf nodes are consistent.

Example 6.4 In the example of Figure 6.2a, $\mathcal{T}$ is the tree rooted at X, and accordingly the root OR node of the AND/OR tree in 6.2c is X. Its child nodes which are AND nodes, are labeled $\langle X, 1\rangle, \langle X, 2\rangle, \langle X, 3\rangle$ (only the values are noted in the figure). From each of these AND nodes emanate two OR nodes, Y and Z, since these are the child nodes of X in the guiding tree of Figure 6.2a. The descendants of Y along the path from the root, $\langle X, 1\rangle$, are $\langle Y, 2\rangle$ and $\langle Y, 3\rangle$ only, since $\langle Y, 1\rangle$ is inconsistent with $\langle X, 1\rangle$. In the next level, from each node $\langle Y, y\rangle$ emanate OR nodes labeled T and R and from $\langle Z, z\rangle$ emanate nodes labeled L and M as dictated by the guiding tree. In Figure 6.2c a solution tree is highlighted.

As noted, if the graphical model is not a tree it can be guided by some legal spanning tree of the graph. For example, as we will show in Section 6.1.2, a depth-first-search (DFS) spanning tree of the graph is a useful and legal guiding tree. The notion of a DFS spanning tree is defined for undirected graphs.

Definition 6.5 DFS spanning tree. Given a graph $G = (V, E)$ and given a node X_1, a *DFS* tree $\mathcal{T}$ of G is generated by applying a depth-first-search traversal over the graph, yielding an ordering $d = (X_1, \ldots, X_n)$. The *DFS spanning tree* $\mathcal{T}$ of G is defined as the tree rooted at the first node, X_1, and which includes only the traversed (by DFS) arcs of G. Namely, $\mathcal{T} = (V, E')$, where $E' = \{(X_i, X_j) \mid X_j \ traversed \ from \ X_i \ by \ DFS \ traversal\}$.

Example 6.6 Consider the probabilistic network given in Figure 6.3a whose undirected primal graph is obtained by including the broken arcs and removing the arrows. A guiding tree which in this case is a DFS spanning-tree of the graph is given in part (b). The dashed arcs are part of the graph but not the spanning-tree arcs. The AND/OR search tree associated with this guiding tree is given in part (d) of the figure. The weights on the arcs will be explained next.

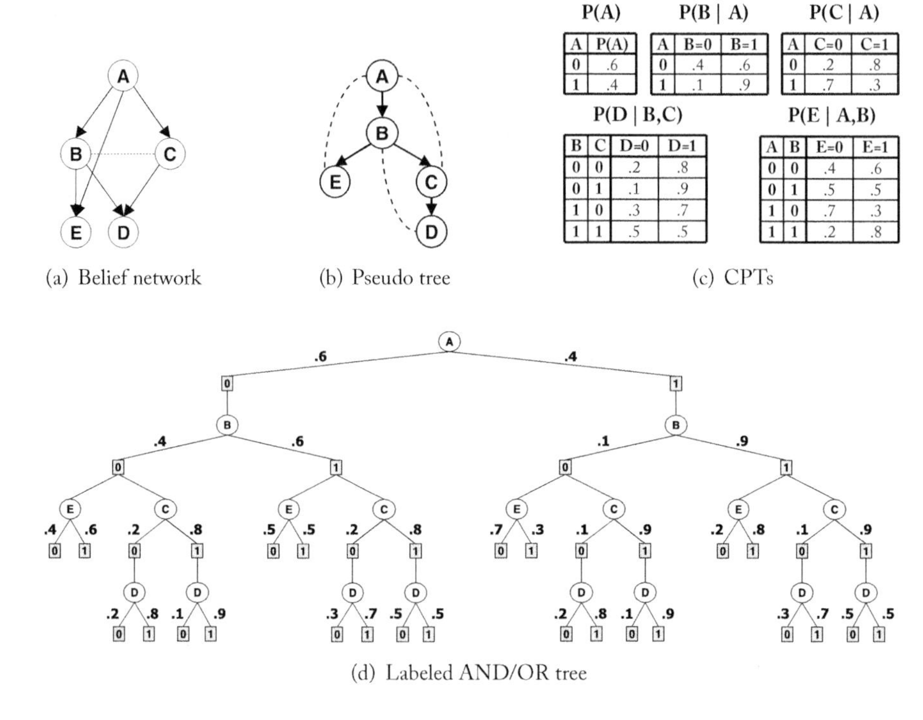

P(A)

A	P(A)
0	.6
1	.4

P(B | A)

A	B=0	B=1
0	.4	.6
1	.1	.9

P(C | A)

A	C=0	C=1
0	.2	.8
1	.7	.3

P(D | B,C)

B	C	D=0	D=1
0	0	.2	.8
0	1	.1	.9
1	0	.3	.7
1	1	.5	.5

P(E | A,B)

A	B	E=0	E=1
0	0	.4	.6
0	1	.5	.5
1	0	.7	.3
1	1	.2	.8

Figure 6.3: Labeled AND/OR search tree for belief networks.

6.1.1 WEIGHTS OF OR-AND ARCS

The arcs in AND/OR trees are associated with weights defined based on the graphical model's functions and the combination operator. The simplest case is that of constraint networks.

Definition 6.7 arc weights for constraint networks. In an AND/OR search tree $S_{\mathcal{T}}(\mathcal{R})$ of a constraint network $\mathcal{R}$, each terminal node is assumed to have a single, dummy, outgoing arc. The outgoing arc of a terminal AND node always has the weight "1" (namely it is consistent). An outgoing arc of a terminal OR node has weight "0", (there is no consistent value assignments if an OR node is a leaf). The weight of any internal OR to AND arc is "1." The arcs from AND to OR nodes have no weight.

We next define arc weights for any general graphical model using the notion of buckets of functions. The concept is simple even if the formal definition may look complex. When considering an arc (n, m) having labels (X_i, x_i) (X_i labels n and x_i labels m), we identify all the functions over variable X_i that are fully instantiated in path(n) once X_i is assigned. We then associate each function with its valuation given the current value-assignment along the path to n. The products

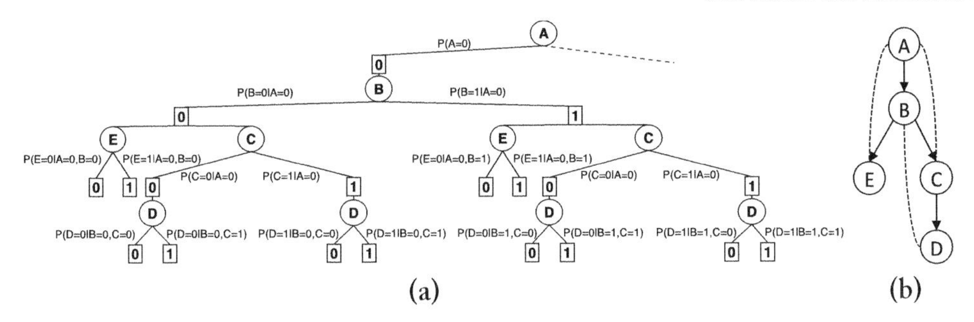

Figure 6.4: Arc weights for probabilistic networks.

of all these *function values* is the weight of the arc. The following definition identify those functions of X_i we want to consider in the product. Weights are assigned only on arcs connecting an OR node to an AND node.

Definition 6.8 OR-to-AND weights, buckets relative to a tree. Given an AND/OR tree $S_{\mathcal{T}}(\mathcal{M})$, of a graphical model $\mathcal{M}$, the weight $w_{(n,m)}(X_i, x_i)$ of arc (n, m) is the *combination* (e.g., product) of all the functions in X_i's bucket relative to $\mathcal{T}$, denoted $B_{\mathcal{T}}(X_i)$, which are assigned by their values along $path(m)$. $B_{\mathcal{T}}(X_i)$ include all functions f having X_i in their scopes and whose $scope(f) \subseteq path_{\mathcal{T}}(X_i)$. Formally, $w_{(n,m)}(X_i, x_i) = \bigotimes_{f \in B_{\mathcal{T}}(X_i)} f(val(path(m)))$. If the set of functions is empty the weight is the constant 1 (or the identity relative to the combination operator).

Definition 6.9 Weight of a solution subtree. Given a weighted AND/OR tree $S_{\mathcal{T}}(\mathcal{M})$, of a graphical model $\mathcal{M}$, the weight of a subtree t is $w(t) = \bigotimes_{e \in arcs(t)} w(e)$, where $arcs(t)$ is the set of arcs in subtree t.

Example 6.10 Figure 6.4 shows a guiding DFS tree of the Bayesian network in Figure 6.3, along a guiding tree in 6.4b and a portion of the AND/OR search tree with the appropriate weights on the arcs expressed symbolically. The bucket of variable E contains the function $P(E|A, B)$, and the bucket of C contains two functions, $P(C|A)$ and $P(D|B, C)$. Note that $P(D|B, C)$ belongs neither to the bucket of B nor to the bucket of D, but it is contained in the bucket of C, which is the last variable in its scope to be instantiated in a path from the root of the tree. We see indeed that the weights from nodes labeled E and from any of its AND value assignments include only the instantiated function $P(E|A, B)$, while the weights on the arcs connecting C to its AND child nodes are the products of the two functions in its bucket, instantiated appropriately. (Exercise: show how would the weight computed on the arc would change if we actually use the guiding tree in Figure 6.3b). The evaluated weights along this pseudo-tree are depicted in Figure 6.3d.

6.1.2 PSEUDO TREES

We have mentioned that a *DFS* spanning tree is a legal guiding tree for the AND/OR search. This is indeed the case because child nodes branching reflect problem decomposition. However, there is a more general class of spanning trees, called *pseudo-trees*, which can be considered. In order to guide a proper decomposition for a graphical model, such trees need to obey the back-arc property.

Definition 6.11 Pseudo tree, extended graph. Given an undirected graph $G = (V, E)$, a directed rooted tree $\mathcal{T} = (V, E')$ defined on all its nodes is a *pseudo tree* if any arc in E which is not in E' is a back-arc in $\mathcal{T}$, namely, it connects a node in $\mathcal{T}$ to an ancestor in $\mathcal{T}$. The arcs in E' may not all be included in E. Given a pseudo tree $\mathcal{T}$ of G, the *extended graph* of G relative to $\mathcal{T}$ includes also the arcs in E' that are not in E. That is, the extended graph is defined as $G^{\mathcal{T}} = (V, E \cup E')$.

Clearly, a DFS-tree is a pseudo-tree with the additional restriction that all its arcs are in included in the original graph. The use of a larger class of pseudo trees has the potential of yielding smaller depth guiding trees which are highly desirable, as we show in the next example.

Example 6.12 Consider the graph G displayed in Figure 6.5a. Ordering $d_1 = (1, 2, 3, 4, 7, 5, 6)$ is a DFS ordering of a DFS spanning tree $\mathcal{T}_1$ having depth of 3 (Figure 6.5b). The tree $\mathcal{T}_2$ in Figure 6.5c is a pseudo tree and has a tree depth of 2 only. The two tree-arcs (1,3) and (1,5) are not in G. The tree $\mathcal{T}_3$ in Figure 6.5d, is a chain. The extended graphs $G^{\mathcal{T}_1}$, $G^{\mathcal{T}_2}$ and $G^{\mathcal{T}_3}$ are presented in Figure 6.5b, c, d when we ignore directionality and include the broken arcs.

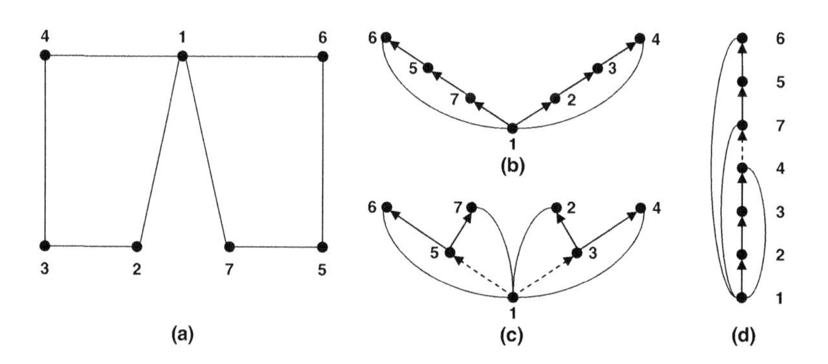

Figure 6.5: (a) A graph; (b) a DFS tree $\mathcal{T}_1$; (c) a pseudo tree $\mathcal{T}_2$; and (d) a chain pseudo tree $\mathcal{T}_3$.

Figure 6.6 shows the AND/OR search trees along the pseudo trees $\mathcal{T}_1$ and $\mathcal{T}_2$ in Figure 6.5. The domains of the variables are $\{a, b, c\}$ and there is no pruning due to hard constraints. We see that the AND/OR search tree based on $\mathcal{T}_2$ is smaller because $\mathcal{T}_2$ has a smaller height than $\mathcal{T}_1$. The weights are not specified here.

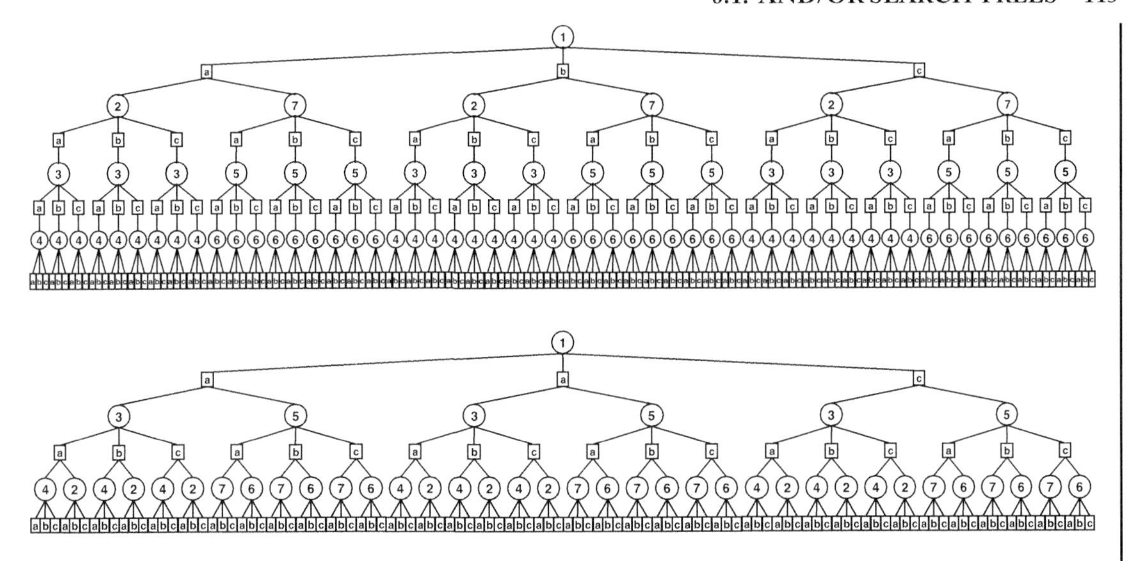

Figure 6.6: AND/OR search tree along pseudo trees $\mathcal{T}_1$ and $\mathcal{T}_2$.

6.1.3 PROPERTIES OF AND/OR SEARCH TREES

Any pseudo tree $\mathcal{T}$ of a graph G has the property that the arcs of G which are not in $\mathcal{T}$ are back arcs. Namely, they connect a node to one of its ancestors in the guiding tree. This property implies that each scope of a function in F will be fully assigned on some path in $\mathcal{T}$, a property that is essential for the ability of the AND/OR search space to consider all the functions in the model and supports correct computation. In fact, the AND/OR search tree can be viewed as an alternative representation of the graphical model.

Theorem 6.13 Correctness. *Given a graphical model $\mathcal{M} = \langle \mathbf{X}, \mathbf{D}, \mathbf{F} = \{f_1, ..., f_r\}, \bigotimes \rangle$ having a primal graph G and a guiding pseudo-tree $\mathcal{T}$ of G and its associated weighted AND/OR search tree $S_{\mathcal{T}}(\mathcal{M})$ then (1) there is a one-to-one correspondence between solution subtrees of $S_{\mathcal{T}}(\mathcal{M})$ and solutions of $\mathcal{M}$; (2) the weight of any solution tree equals the cost of the full solution assignment it denotes; namely, if t is a solution tree of $S_{\mathcal{T}}(\mathcal{M})$ then $F(val(t)) = w(t)$, where $val(t)$ is the full solution defined by tree t. (See Appendix for a proof.)*

As already mentioned, the virtue of an AND/OR search tree representation is that its size can be far smaller than the traditional OR search tree. The size of an AND/OR search tree depends on its depth, also called height, of its pseudo-tree $\mathcal{T}$. Therefore, pseudo trees of smaller height should be preferred. An AND/OR search tree becomes an OR search tree when its pseudo tree is a chain.

Theorem 6.14 Size of AND/OR search tree. *Given a graphical model $\mathcal{M}$, with domains size bounded by k, having a pseudo tree $\mathcal{T}$ whose height is h and having l leaves, the size of its AND/OR*

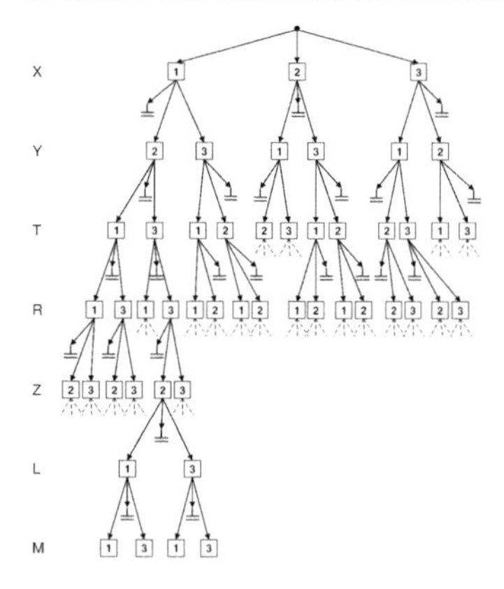

Figure 6.7: OR search tree for the tree problem in Figure 6.2a.

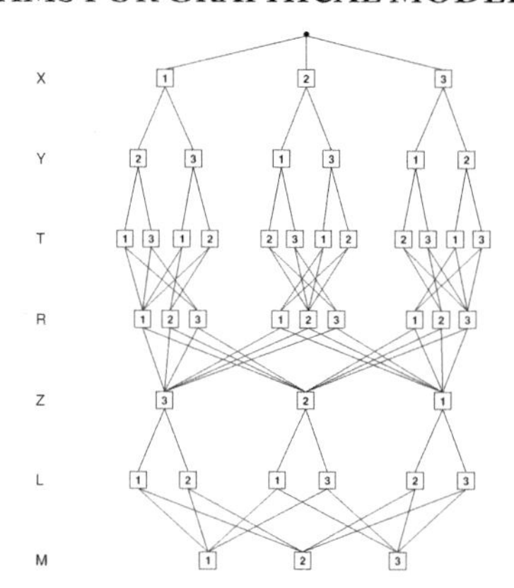

Figure 6.8: The minimal OR search graph of the tree graphical model in Figure 6.2a.

search tree $S_{\mathcal{T}}(\mathcal{M})$ is $O(l \cdot k^h)$ and therefore also $O(nk^h)$ and $O((bk)^h)$ when b bounds the branching degree of $\mathcal{T}$ and n bounds the number of nodes. The size of its OR search tree along any ordering is $O(k^n)$ and these bounds are tight. (See Appendix for proof.)

We can give a more refined bound on the search space size by spelling out the height h_i of each leaf L_i in $\mathcal{T}$ as follows. Given a guiding spanning $\mathcal{T}$ having $L = \{L_1, \ldots, L_l\}$ leaves of a model $\mathcal{M}$, where the depth of leaf L_i is h_i and k bounds the domain sizes, the size of its full AND/OR search tree $S_{\mathcal{T}}(\mathcal{M})$ is $O(\sum_{k=1}^{l} k^{h_i+1})$. Using also the domain sizes for each variable yields an even more accurate expression of the search tree size: $|S_{\mathcal{T}}(\mathcal{M})| = O(\sum_{L_k \in L} \prod_{\{X_j | X_j \in path_{\mathcal{T}}(L_k)\}} |D(X_j)|)$.

6.2 AND/OR SEARCH GRAPHS

It is often the case that a search space that is a tree can become a graph if nodes that root identical search subspaces, or correspond to identical subproblems, are identified. Any two such nodes can be *merged*, yielding a graph and thus reducing the size of the search space.

Example 6.15 Consider again the graph in Figure 6.2a and its AND/OR search tree in Figure 6.2c depicted again in Figure 6.9 representing a constraint network. Observe that at level 3, node $\langle Y, 1\rangle$ appears twice, (and so are $\langle Y, 2\rangle$ and $\langle Y, 3\rangle$) (not shown explicitly in the figure). Clearly however, the subtrees rooted at each of these two AND nodes are identical and they can be merged because in this tree model, any specific assignment to Y uniquely determines its rooted subtree. Indeed, the resulting merged AND/OR search graph depicted in Figure 6.10 is equivalent to the AND/OR search tree in Figure 6.9.

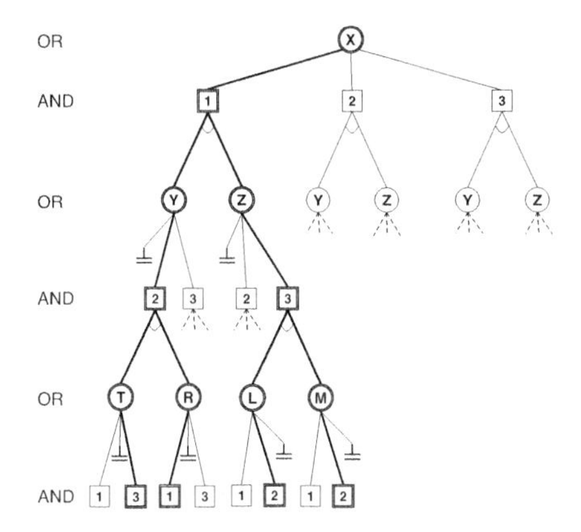

Figure 6.9: AND/OR search tree for the tree problem in Figure 6.2a.

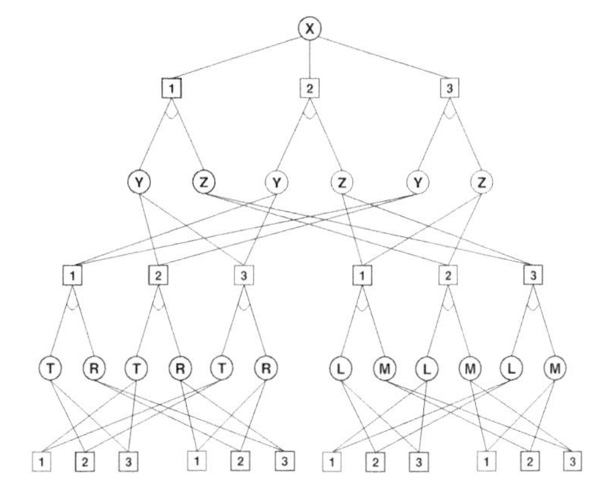

Figure 6.10: The minimal AND/OR search graph of the tree graphical model in Figure 6.2a.

It may also occur that two nodes that do not root identical subtrees still correspond to equivalent subproblems. Such nodes can also be *unified*, even if their *explicit* weighted subtrees do not look identical. To discuss this issue we need the notion of *equivalent graphical models*. In general, two graphical models are equivalent if they have the same set of solutions, and if each is associated with the same *cost*. We will use the notion of *universal graphical model* to define *equivalence*. A universal graphical model represents the *solutions* of a graphical model, through a single global function over all the variables. For example, the universal model of a Bayesian network is the joint probability distribution it represents.

Definition 6.16 Universal equivalent graphical model. Given a graphical model $\mathcal{M} = \langle X, D, F, \bigotimes \rangle$ the universal equivalent model of $\mathcal{M}$ is $u(\mathcal{M}) = \langle X, D, F = \{\bigotimes_{i=1}^{r} f_i\} \rangle$.

We also need to define the cost of a partial solution and the notion of a graphical model conditioned on a partial assignment. Informally, a graphical model conditioned on a particular partial assignment is obtained by assigning the appropriate values to all the relevant variables in the function (to all the conditioning set) and modifying the output functions appropriately.

Definition 6.17 Cost of an assignment, conditional model. Given a graphical model $\mathcal{R} = \langle \mathbf{X}, \mathbf{D}, \mathbf{C}, \bigotimes \rangle$:

1. the cost of a full assignment $\mathbf{x} = (x_1, ..., x_n)$ is defined by $c(\mathbf{x}) = \bigotimes_{f \in F} f(x_f)$. The *cost of a partial assignment* $\mathbf{y}$, over $\mathbf{Y} \subseteq \mathbf{X}$ is the combination of all the functions whose scopes are included in $\mathbf{Y}$ (denoted F_Y) evaluated at the assigned values. Namely, $c(y) = \bigotimes_{f \in F_{\mathbf{Y}}} f(y_f)$.
2. the graphical model conditioned on $Y = y$ is $\mathcal{M}|_y = \langle X - Y, D|_{X-Y}, F|_y, \bigotimes \rangle$, where $F|_y = \{f|_{Y=y}, f \in F\}$.

6.2.1 GENERATING COMPACT AND/OR SEARCH SPACES

We will next define the merge operator. It transforms AND/OR search trees into an equivalent AND/OR graphs.

Definition 6.18 Merge. Assume a given weighted AND/OR search graph $S'_{\mathcal{T}}$ of a graphical model $\mathcal{M}$ and assume two paths $path(n_1)$ and $path(n_2)$ ending by AND nodes at level i having the same label x_i. Nodes n_1 and n_2 can be *merged* iff the weighted search subgraphs rooted at n_1 and n_2 are identical. The *merge* operator, $merge(n_1, n_2)$, redirects all the arcs going into n_2 into n_1 and removes n_2 and its subgraph. When we merge AND nodes only we call the operation AND-merge. The same reasoning can be applied to OR nodes, and we call the operation OR-merge.

Proposition 6.19 Merge-minimal AND/OR graphs *Given a weighted AND/OR search graph $\mathcal{G}_{\mathcal{T}}$ guided by a pseudo tree $\mathcal{T}$: The* merge *operator has a unique fix point, called the* **merge-minimal** *AND/OR search graph. (See proof in the Appendix).*

When $\mathcal{T}$ is a chain pseudo tree, the above definitions are applicable to the traditional OR search tree as well. However, we may not be able to reach the same compression as in some AND/OR cases, because of the linear structure imposed by the OR search tree.

Example 6.20 The smallest OR search graph of the graph-coloring problem in Figure 6.2a (depicted again in Figure 6.7) is given in Figure 6.8 along the DFS order X, Y, T, R, Z, L, M. The smallest AND/OR graph of the same problem along the DFS tree is given in Figure 6.10. We see that some variable-value pairs (AND nodes) must be repeated in Figure 6.8 while in the AND/OR case (Figure 6.10) they appear just once. In particular, the subgraph below the paths $(\langle X, 1\rangle, \langle Y, 2\rangle)$ and $(\langle X, 3\rangle, \langle Y, 2\rangle)$ in the OR tree cannot be merged at $\langle Y, 2\rangle$. You can now compare all the four search space representations side by side in Figures 6.7–6.10.

6.2.2 BUILDING CONTEXT-MINIMAL AND/OR SEARCH GRAPHS

The merging rule seems to be quite operational; we can generate the AND/OR search tree and then recursively merge identical subtrees going from leaves to root. This, however, requires generating the whole search tree first, which would still be costly. It turns out that for some nodes it is possible to recognize that they can be merged by using graph properties only, namely based on their *contexts*. The context of a variable is the set of its ancestor variables in the pseudo tree $\mathcal{T}$ that completely determine the conditioned subproblems below it.

We have already seen in Figure 6.2a that at level 3, node $\langle Y, 1\rangle$ appears twice (and so are $\langle Y, 2\rangle$ and $\langle Y, 3\rangle$). Clearly we can see that Y uniquely determines its rooted subtree. In this case Y is its own context and the AND/OR search graph in Figure 6.10 is equivalent to the AND/OR search tree in Figure 6.7. In general, an AND/OR search graph of a graphical model that is a tree can be obtained by merging all AND nodes having the same label $\langle X, x\rangle$. That is, every variable is its own context. The resulting equivalent AND/OR graph has search space size of $O(nk)$.

The general idea of a context is to identify a minimal set of ancestor variables, along the path from the root to the node in the pseudo tree, such that when assigned the same values they yield the same conditioned subproblem, regardless of value assigned to the other ancestors. To derive a general merging scheme we define the induced-width of a pseudo-tree.

Definition 6.21 Induced width of a pseudo tree. The induced width of G relative to a pseudo tree $\mathcal{T}$, is the maximum width of its *induced pseudo tree* obtained by recursively connecting the parents of each node, going from leaves to root along each branch. In that process we consider both the extended arcs in the pseudo tree and those in the graphical model.

Definition 6.22 Parents, parents-separators. Given a primal graph G and a pseudo tree $\mathcal{T}$ of its graphical model $\mathcal{M} = \langle \mathbf{X}, \mathbf{D}, \mathbf{C}, \bigotimes\rangle$, the *parents* of an OR node X_i, denoted by pa_i or pa_{X_i}, are the ancestors of X_i that are connected in G to X_i or to descendants of X_i. The *parent-separators* of X_i (or of $\langle X_i, x_i\rangle$), denoted by pas_{X_i}, are formed by X_i and its ancestors that have connections in G to descendants of X_i.

It follows from these definitions that the parents of X_i, pa_{X_i} separate in the primal graph G, the ancestors of X_i in $\mathcal{T}$, from X_i and its descendants. Similarly, the parents-separators of X_i, pas_{X_i}, separate the ancestors of X_i from its descendants. It is also easy to see that each variable X_i and its parents pa_{X_i} form a clique in the induced pseudo-graph. The following proposition establishes the relationship between pa_{X_i} and pas_{X_i}. We use both in order to characterize two types of merging: AND merge and OR merge. The following claim follows directly from Definitions 6.22. It is easy to see the following.

Proposition 6.23 Relations between contexts

1. *If Y is the single child of X in $\mathcal{T}$, then $pas_X = pa_Y$.*
2. *If X has children $Y_1, \ldots, Y_k$ in $\mathcal{T}$, then $pas_X = \cup_{i=1}^{k} pa_{Y_i}$.*

Theorem 6.24 Context-based merge operators. *Let $G^{\mathcal{T}^*}$ be the induced pseudo tree of $\mathcal{T}$ and let $path(n_1)$ and $path(n_2)$ be any two partial paths in an AND/OR search graph.*

1. *If n_1 and n_2 are AND nodes annotated by $\langle X_i, x_i\rangle$ and*

$$val(path(n_1))[pas_{X_i}] = val(path(n_2))[pas_{X_i}] \quad (6.2)$$

then the AND/OR search subtrees rooted by n_1 and n_2 can be merged.

2. *If n_1 and n_2 are OR nodes annotated by X_i and*

$$val(path(n_1))[pa_{X_i}] = val(path(n_2))[pa_{X_i}] \tag{6.3}$$

then the AND/OR search subtrees rooted by n_1 and n_2 can be merged.

Definition 6.25 context. The $val(path(n_i))[pas_{X_i}]$ is called the **AND context** of n_i and the $val(path(n_i))[pa_{X_i}]$ is called the **OR context** of n_i.

Example 6.26 For the balanced tree in Figure 6.2a consider the *chain* pseudo tree $d = (X, Y, T, R, Z, L, M)$. Namely, the chain has arcs $\{(X, Y), (Y, T), (T, R), (R, Z), (Z, L), (L, M)\}$ and the extended graph includes also the arcs $(Z, X), (M, Z)$. The context of T along d is XYT (since the induced graph has the arc (T, X)), of R it is XR, for Z it is Z and for M it is M. Indeed in the first three levels of the OR search graph in Figure 6.8 there are no merged nodes. In contrast, if we consider the AND/OR ordering along the DFS tree, the context of every node is itself yielding a single appearance of each AND node having the same assignment annotation in the minimal AND/OR graph (See Figure 6.10 and contrast it with Figure 6.8).

Definition 6.27 Context minimal AND/OR search graph. The AND/OR search graph of $\mathcal{M}$ guided by a pseudo-tree $\mathcal{T}$ that is closed under context-based merge operator, (namely no more merging is possible), is called the *context minimal* AND/OR search graph and is denoted by $CM_{\mathcal{T}}(\mathcal{R})$.

We should note that we can, in general, merge nodes based both on AND and OR contexts. However, Proposition 6.23 shows that doing just one type of merging renders the other unnecessary (up to some small constant factor). In practice, we would recommend just the OR context based merging, because it has a slight (albeit by a small constant factor) space advantage.

Example 6.28 Figure 6.11a refer back to the model given in Figure 6.3a, again assuming that all assignments are valid and that variables take binary values. Figure 6.11b shows, again, the pseudo tree derived from ordering $d = (A, B, E, C, D)$. The (OR) context of each node appears in square brackets, and the broken arcs are backarcs. The context-minimal AND/OR graph appears in 6.11(b).

Since each context must appear only once in the Context-minimal graph (different appearances should be merged) the number of nodes in the context minimal AND/OR search graph cannot exceed the number of different contexts. Since, as we will show, the context's scope size is

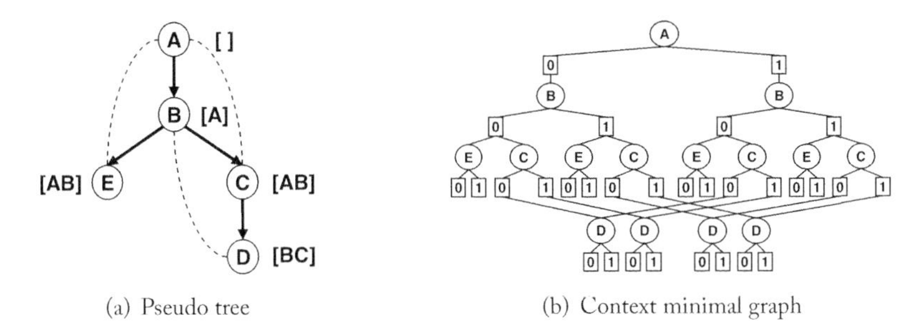

(a) Pseudo tree

(b) Context minimal graph

Figure 6.11: AND/OR search graph.

bounded by the induced width of the pseudo tree that guides it, the size of the context minimal graph can be bounded exponentially by the induced width along the pseudo-tree.

Proposition 6.29 *Given a graphical model $\mathcal{M}$, and a pseudo tree $\mathcal{T}$ having induced width w, then the size of the context minimal AND/OR search graph based on $\mathcal{T}$, $CM_{\mathcal{T}}(\mathcal{R})$, is $O(n \cdot k^w)$, when k bounds the domain size and n is the number of variables.*

Proof. For any variable, the number of its contexts is bounded by the number of possible instantiations to the variables in it context. Since the context size of each variable is bounded by its induced-width along the pseudo tree (prove as an exercise), we get the bound of $O(k^w)$. Since we have n variables, the total bound is $O(n \cdot k^w)$. □

In summary, context-based merge (AND and/or OR) offers a powerful way of trimming the size of the AND/OR search space, and therefore of bounding the truly minimal AND/OR search graph. We can generate $CM_{\mathcal{T}}$ using depth-first or breadth first traversals while figuring the converging arcs into nodes via their contexts. This way we avoid generating duplicate searches for the same contexts. All in all, the generation of the search graph is linear in its size. The AND/OR graph is exponential in w and linear in n. Based on Proposition 6.29 we can conclude the following.

Theorem 6.30 *The context minimal AND/OR search graph $CM_{\mathcal{T}}$ of a graphical model whose guiding pseudo tree has a treewidth w can be generated in time and space $O(nk^{w+1})$ and $O(nk^w)$, respectively. (Prove as a exercise.)*

6.3 FINDING GOOD PSEUDO TREES

Since the AND/OR search space, be it a tree or a graph, depends on a guiding pseudo-tree we should address the issue of finding good pseudo-tress. We will discuss two schemes for generating good pseudo-trees. One based on an induced graph along an ordering of the variables, while the other is based on hypergraph-decomposition.

6.3.1 PSEUDO TREES CREATED FROM INDUCED GRAPHS

We saw that the complexity of an AND/OR search trees is controlled by the height of the pseudo tree. It is desirable therefore to find pseudo trees having minimal height. This is yet another graph problem (in addition to finding minimal induced-width) which is known to be NP-complete but greedy algorithms and polynomial time heuristic scheme are available.

A general scheme for generating pseudo trees starts from an induced graphs along some ordering d. A pseudo-tree can then be obtained via a depth-first traversal of the induced-ordered graph starting from the first node in d and breaking ties in favor of earlier variables in d. An alternative way for generating a pseudo-tree from an induced ordered graph is based on the observation that a bucket tree is a pseudo tree (see Definition 5.2). Summarizing:

Proposition 6.31 *Given a graphical model* $\mathcal{M} =< X, D, F, \bigotimes >$ *and an ordering* d,

1. *the bucket tree derived from the induced ordered graph along* d *of* $\mathcal{M}$, $T = (X, E)$ *with* $E = \{(X_i, X_j)|(B_{X_i}, B_{X_j}) \in bucket - tree\}$, *is a pseudo tree of* $\mathcal{M}$, *and*

2. *the DFS-tree generated by traversing the induced-order graph starting at the first variable of its ordering, is a pseudo tree.*

Proof. All one need to show is that all the arcs in the primal graph of $\mathcal{M}$ which are not in T are back-arcs and this is easy to verify based on the construction of DFS tree in part (1) and of a bucket-tree in part (2). (Exercise: complete the proof). □

It is interesting to note that a chain graphical model has a (non-chain) pseduo-tree of depth $\log n$, when n is the number of variables. The induced width of such a tree is $\log n$ as well. On the other hand the minimum induced width of a chain pseudo tree is 1. Therefore, on the one hand a chain can be solved in linear space and in $O(k^{\log n})$ time along its $log n$ height pseudo tree, and on the other hand it can also be solved in $O(nk^2)$ time with $O(nk)$ memory using bucket-elimination along its chain whose induced width is 1 and height is $n/2$. This example generalizes into a relationship between the treewidth and the pseudo-tree height of a graph.

Proposition 6.32 *[Bayardo and Miranker, 1996; H.L. Bodlaender and Kloks, 1991] The minimal height,* h^*, *of all pseudo trees of a given graph* G *satisfies* $h^* \leq w^* \cdot \log n$, *where* w^* *is the tree width of* G.

Table 6.1: Bayesian networks repository (left); SPOT5 benchmarks (right).

Network	hypergraph		min-fill		Network	hypergraph		min-fill	
	width	depth	width	depth		width	depth	width	depth
barley	7	13	7	23	spot_5	47	152	39	204
diabetes	7	16	4	77	spot_28	108	138	79	199
link	21	40	15	53	spot_29	16	23	14	42
mildew	5	9	4	13	spot_42	36	48	33	87
munin1	12	17	12	29	spot_54	12	16	11	33
munin2	9	16	9	32	spot_404	19	26	19	42
munin3	9	15	9	30	spot_408	47	52	35	97
munin4	9	18	9	30	spot_503	11	20	9	39
water	11	16	10	15	spot_505	29	42	23	74
pigs	11	20	11	26	spot_507	70	122	59	160

Proof. If there is a tree decomposition of G having a treewidth w, then we can create a pseudo tree whose height h satisfies $h \leq w \cdot \log n$ (prove as an exercise). From this it follows that $h^* \leq w^* \cdot \log n$. □

The above relationship suggests a bound on the size of AND/OR search trees of a graphical models in terms of their treewidth.

Theorem 6.33 *A graphical model that has a treewidth w^* has an AND/OR search tree whose size is $O(k^{(w^* \cdot \log n)})$, where k bounds the domain size and n is the number of variables.*

Notice, however, that even though a graph may have induced-width of w^*, the induced width of the pseudo tree created as suggested by the above theorem may be of size $w^* logn$ and not w^*.

Width vs. height of a given pseudo tree. Since an induced-ordered graph can be a starting point in generating a pseudo tree, the question is if the min-fill ordering heuristic which appears to be quite good for finding small induced-width is also good for finding pseudo trees with small heights (see Chapter 3). A different question is what is the relative impact of the width and the height on the actual search complexity. The AND/OR search graph is bounded exponentially by the induced-width while the AND/OR search tree is bounded exponentially by the height. We will have a glimpse into these questions by comparing with an alternative scheme for generating pseudo trees which is based on the hypergraph decompositions scheme.

6.3.2 HYPERGRAPH DECOMPOSITIONS

Definition 6.34 Hypergraph separators. Given a dual hypergraph $\mathcal{H} = (\mathbf{V}, \mathbf{E})$ of a graphical model, a *hypergraph separator decomposition* of size k by nodes S is obtained if removing S yields a hypergaph having k disconnected components. S is called a separator.

It is well known that the problem of finding the minimal size hypergraph separator is hard. However, heuristic approaches were developed over the years.[1]. Generating a pseudo tree $\mathcal{T}$ (yielding also a tree-decomposition) for $\mathcal{M}$ using hypergraph decomposition is fairly straightforward. The vertices of the hypergraph are partitioned into two balanced (roughly equal-sized) parts, denoted by $\mathcal{H}_{left}$ and $\mathcal{H}_{right}$, respectively, while minimizing the number of hyperedges across. A small number of crossing edges translates into a small number of variables shared between the two sets of functions. $\mathcal{H}_{left}$ and $\mathcal{H}_{right}$ are then each recursively partitioned in the same fashion, until they contain a single vertex. The result of this process is a tree of hypergraph separators which can be shown to also be a pseudo tree of the original model where each separator corresponds to a subset of variables connected by a chain.

Table 6.1 illustrates and contrasts the induced width and height of pseudo trees obtained with the hypergraph and min-fill heuristics for 10 Bayesian networks from the Bayesian Networks Repository[2] and 10 constraint networks derived from the SPOT5 benchmarks [Bensana *et al.*, 1999]. It is generally observed that the min-fill heuristic generates lower induced width pseudo trees, while the hypergraph heuristic produces much smaller height pseudo trees. Note that it is not possible to generate a pseudo-tree that is optimal w.r.t. both the treewidth and the height (remember our earlier example of a chain).

Notice that for graphical models having a bounded treewidth w, the minimal AND/OR graph is bounded by $O(nk^w)$ while the minimal OR graph is bounded by $O(nk^{w \cdot \log n})$. We conclude this section with the following example which is particularly illustrative of the tradeoff involved.

Example 6.35 Consider the graph of a graphical model given in Figure 6.12a. We see the pseudo tree in part (b) having w=4 and h=8 and the corresponding context-minimal search graph in (c). The second pseudo-tree in part (d) has w=5, h=6 and the context-minimal graph appears in part (e).

6.4 VALUE FUNCTIONS OF REASONING PROBLEMS

As we described earlier, there are a variety of reasoning problems over graphical models (see Chapter 2). For constraint networks, the most popular tasks are to decide if the problem is consistent,

[1]A good package hMeTiS is Available at: http://www-users.cs.umn.edu/karypis/metis/hmetis
[2]Available at: `http://www.cs.huji.ac.il/labs/compbio/Repository`

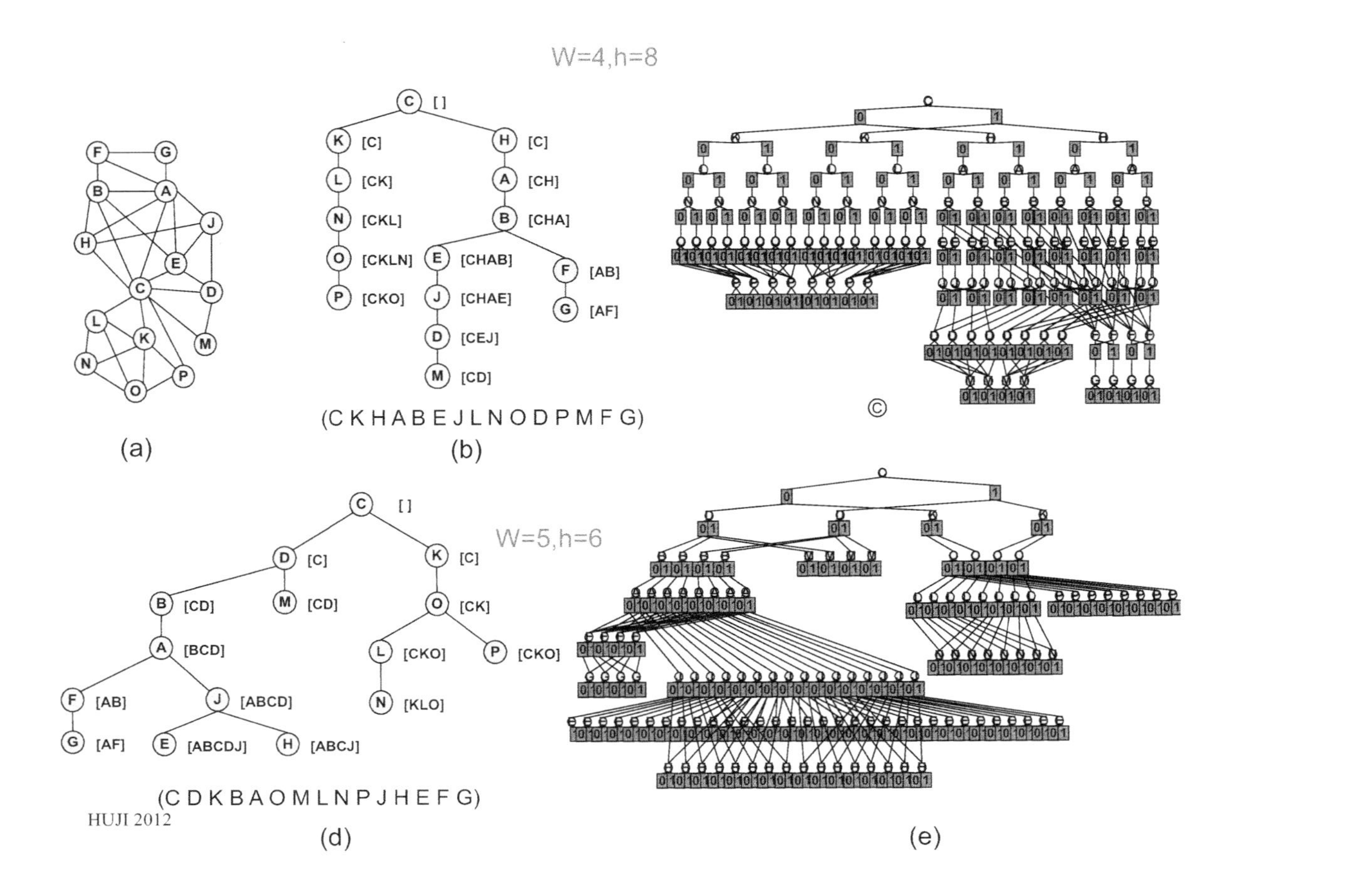

Figure 6.12: A graphical model; (a) one pseudo-tree; (b) its context-minimal search graph; (c) a second pseudo-tree; (d) its corresponding context-minimal AND/OR search graph; and (e) its corresponding context-minimal AND/OR search graph

to find a single solution or to count solutions. If a cost function is defined by the graphical model we may also seek an optimal solution. The primary tasks over probabilistic networks are computing beliefs (i.e., the posterior marginals given evidence), finding the probability of the evidence and finding the most likely tuple given the evidence (i.e., *mpe* and *map* queries). Each of these reasoning problems can be expressed as finding the *value* of nodes in the *weighted AND/OR search space*.

For example, for the task of finding a solution to a constraint network, the value of every node is either "1" or "0." The value "1" means that the subtree rooted at the node is consistent and "0" otherwise. Therefore, the value of the root node determines the consistency query for the full constraint network. For solutions-counting the value function of each node is the number of solutions of the subproblem rooted at that node.

Given a graphical model and a specification by an AND/OR search space of the model. The submodel associated with a node n in the search space, is the submodel conditioned on all the value assignments along the path from the root node to n.

Definition 6.36 Value function for consistency and counting. Given AND/OR tree $S_T(\mathcal{R})$ of a constraint network. The value of a node (AND or OR) for *deciding consistency* is "1" if it roots a consistent subproblem and "0" otherwise. The value of a node (AND or OR) for *counting solutions* is the number of solutions in the subproblem it roots which is the number of solutions in its subtree.

The value of nodes in the search graph can be expressed as a function of the values of their child nodes, thus allowing a recursive value computation from leaves to root.

Proposition 6.37 Recursive value computation for constraint queries *Consider the following.*

1. *For the consistency task the value of AND leaves is "1" and the value of OR leaves is "0" (they are inconsistent). An internal OR node is labeled "1" if one of its successor nodes is "1" and an internal node has value "1" iff all its child OR nodes have value "1."*

2. *The counting values of leaf AND nodes are "1" and of leaf OR nodes are "0." The counting value of an internal OR node is the sum of the counting-values of all its child nodes. The counting-value of an internal AND node is the product of the counting-values of all its child nodes. (Exercise: prove the proposition.)*

We now move to probabilistic queries. We can generalize to any graphical model and to any query. We provide the recursive definition of values and then prove that it is correct, namely, that it has the intended meaning. Remember that the label of an arc $(X_i, \langle X_i, x_i \rangle)$ along path

$path(n = x_i)$ is defined as $w(X_i, \langle X_i, x_i \rangle) = \prod_{f \in B(X_i)} f(val(path(n = x_i)))$, where $B(X_i)$ are the functions in its bucket (see Definition 6.8.)

Definition 6.38 Recursive value computation for a general reasoning problems. The value function of a reasoning problem $\mathcal{M} = \langle X, D, F, \bigotimes \Downarrow \rangle$ is defined as follows: the value of leaf AND nodes is "1" and of leaf OR nodes is "0." The value of an internal OR node is obtained by *combining* the value of each AND child node with the weight (see Definition 6.8) on its incoming arc and then *marginalizing* over all AND children. The value of an AND node is the combination of the values of its OR children. Formally, if $children(n)$ denotes the children of node n in the AND/OR search graph, then[3]:

$$v(n) = \bigotimes_{n' \in children(n)} v(n'), \qquad \text{if } n = \langle X, x \rangle \text{ is an AND node,}$$
$$v(n) = \Downarrow_{n' \in children(n)} (w_{(n,n')} \bigotimes v(n')), \qquad \text{if } n = X \text{ is an OR node.}$$

Given a reasoning task, the value of the root node is the answer to the query as stated next

Proposition 6.39 *Let $\mathcal{M} = \langle \mathbf{X}, \mathbf{D}, \mathbf{F}, \bigotimes \Downarrow \rangle$ be a graphical model reasoning task and let $n = X_1$ be the root node in an AND/OR search graph $S'_T(\mathcal{M})$. Then the value of X_1 defined recursively by Definition 6.38 obeys that $v(X_1) = \Downarrow_X \bigotimes_{f \in F} f$. (For a formal proof see [Dechter and Mateescu, 2007b].)*

For probabilistic network when the combination is a product and the marginalization is a sum, the value of the root node is the probability of evidence. If we use the max marginalization operator, the value of the root is the *mpe* cost.

Example 6.40 Consider our AND/OR tree example of the probabilistic Bayesian network and assume that we have to find the probability of evidence P(D=1,E=0). The weighted AND/OR tree was depicted in 6.3. In Figure 6.13a we show also the value of each node for the query. Starting at leaf bodes for E and D we see that their values is "0" for the non-evidence value and "1" otherwise, indicated through faded arcs. Therefore the value of OR nodes D is as dictated by the appropriate weights. We see that when we go from leaves to root, the value of an OR node is the sum value of their child nodes, each multiplied by the arc-weight. For example, the value $v(n) = 0.88$ in the left part of the tree is obtained from the values of its child nodes, (0.8 and 0.9), each multiplied by their respective weights (0.2,0.8) yielding $v(n) = 0.2 \cdot 0.8 + 0.8 \cdot 0.9 = 0.88$. The value of AND nodes is the product of value of their OR child nodes. In part (b) we see the value associated with nodes in the AND/OR graph. In this case merging occur only in OR nodes labeled D, and the value of each node is computed in the same way.

[3]We abuse notations here as $\bigotimes$ is defined between matrices or tables and here we have scalars

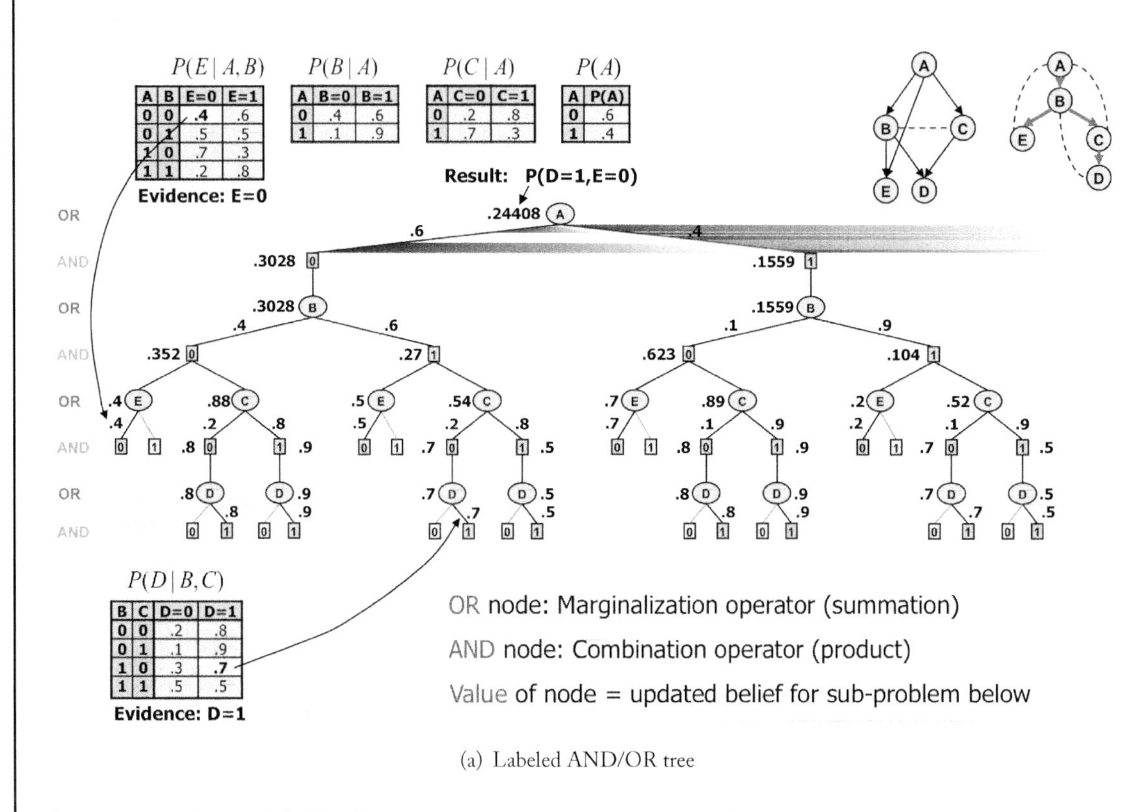

(a) Labeled AND/OR tree

Figure 6.13: Labeled AND/OR search tree and graphs for belief networks.

6.4.1 SEARCHING AND/OR TREE (AOT) AND AND/OR GRAPH (AOG)

Search algorithms that traverse the AND/OR search space can compute the value of the root node yielding the answer to the query. In this section we present a typical depth-first algorithms that traverse AND/OR trees and graphs. We use *solution counting* as an example for a constraint query and the probability of evidence as an example for a probabilistic query. The application of these ideas for combinatorial optimization tasks, such as MPE is straightforward (at least in its brute-force manner). Effective and more sophisticated schemes (e.g., branch and bound or

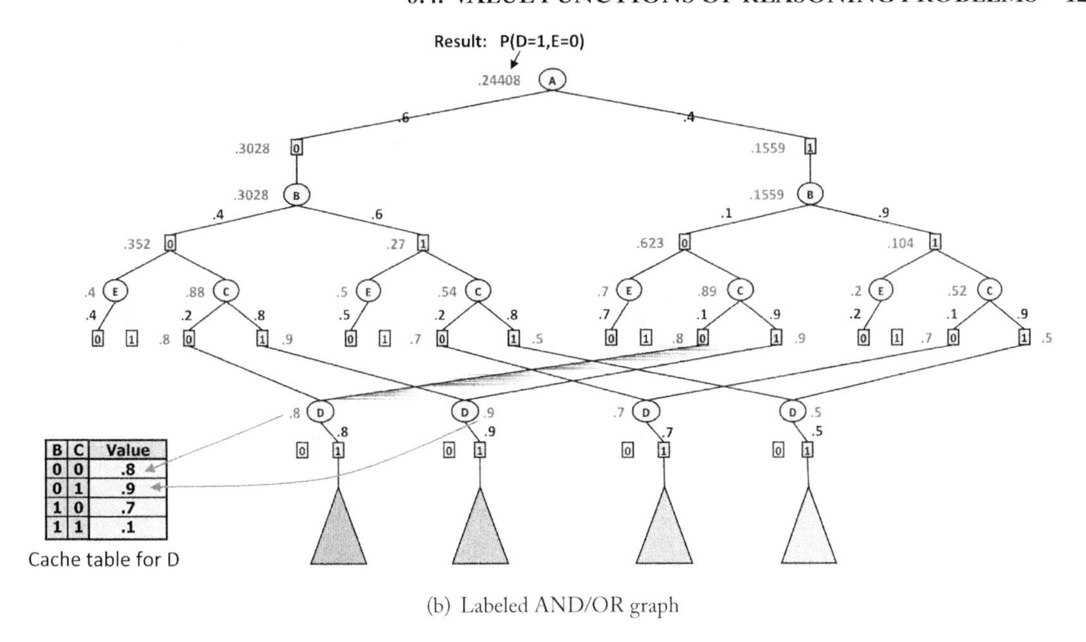

(b) Labeled AND/OR graph

Figure 6.13: Labeled AND/OR search tree and graphs for belief networks.

best-first search) were developed [Marinescu and Dechter, 2005, 2009a,b; Otten and Dechter, 2012].

Example 6.41 Looking again at Figure 6.13 we can see how value computation can be accomplished when we traverse the search space in a depth-first manner, using only linear space. If instead, the AND/OR graph is searched, we need to cache the results of subproblem's value computation in order to avoid redundant computation. This can be accomplished by caching using the node context. For example, in Figure 6.13 we keep a table for node D index by its context consisting of variables B and C. For each path to an OR node labeled D like (A=0, B=0, C=1, D=1), once we discovered that the value below this path is 0.8, we keep this value in the cache table indexed by the pair $(B = 0, C = 1)$. Likewise, for each assignment to these two variables the solution is cached and retrieved when the same context is encountered (see Figure 6.13b).

Algorithm 2, presents the basic depth-first traversal of the AND/OR search tree or search graph for counting the number of solutions of a constraint network, AO-COUNTING, (or for probability of evidence for belief networks, AO-BELIEF-UPDATING). As noted, the context based caching is done using tables. For each variable X_i, a table is reserved in memory for each possible assignment to its parent set pa_i which is its context. Initially each entry has a predefined value, in our case "-1." The fringe of the search is maintained on a stack called `OPEN`. The current node is denoted by `n`, its parent by `p`, and the current path by $path(n)$. The children of the current node are denoted by $successors(\mathtt{n})$. If caching is set to "false" the algorithm searches the AND/OR tree and we will refer to it as AOT.

The algorithm is based on two mutually recursive steps: EXPAND and PROPAGATE, which call each other (or themselves) until the search terminates. Before expanding an OR node, its cache table is checked (line 5). If the same context was encountered before, it is retrieved from cache, and $successors(\mathtt{n})$ is set to the empty set, which will trigger the PROPAGATE step. If a node is not found in cache, it is expanded in the usual way, depending on whether it is an AND or OR node (lines 9–16). The only difference between counting and belief updating is line 11 vs. line 12. For counting, the value of a consistent AND node is initialized to 1 (line 11), while for belief updating, it is initialized to the bucket value for the current assignment (line 12). As long as the current node is not a dead-end and still has unevaluated successors, one of its successors is chosen (which is also the top node on `OPEN`), and the expansion step is repeated.

The bottom up propagation of values is triggered when a node has an empty set of successors (note that as each successor is evaluated, it is removed from the set of successors in line 30). This means that all its children have been evaluated, and its final value can now be computed. If the current node is the root, then the search terminates with its value (line 19). If it is an OR node, its value is saved in cache before propagating it up (line 21). If n is OR, then its parent p is AND and p updates its value by multiplication with the value of n (line 23). If the newly updated value of p is 0 (line 24), then p is a dead-end, and none of its other successors needs to be evaluated. An AND node n propagates its value to its parent p in a similar way, only by summation (line 29). Finally, the current node n is set to its parent p (line 31), because n was completely evaluated. The search continues either with a propagation step (if conditions are met) or with an expansion step.

6.5 GENERAL AND-OR SEARCH - AO(I)

General AND/OR algorithms for evaluating the value of a root node for any reasoning problem using tree or graph AND/OR search are identical to the above algorithms when product is replaced by the combination operator and summation is replaced by the marginalization operator. We can view the AND/OR tree algorithm (which we will denote AOT) and the AND/OR graph algorithm (denoted AOG) as two extreme cases in a parameterized collection of algorithms that trade space for time via a controlling parameter i. We denote this class of algorithms as $AO(i)$ where i determines the size of contexts that the algorithm caches. Algorithm $AO(i)$ records nodes whose context size is i or smaller (the test in line 21 needs to be a bit more elaborate and check if the context size is smaller than i). Thus, AO(0) is identical to AOT, while $AO(w)$ is identical to AOG, where w is the induced width of the used pseudo-tree. For any intermediate i we get an intermediate level of caching, which is space exponential in i and whose execution time will increase as i decreases. Some elaboration follows.

Algorithm 2: AO-COUNTING / AO-BELIEF-UPDATING.

Input: A constraint network $\mathcal{R} = \langle X, D, C \rangle$, or a belief network $\mathcal{B} = \langle X, D, P \rangle$; a pseudo tree $\mathcal{T}$ rooted at X_1; parents pa_i (OR-context) for every variable X_i; caching set to $true$ or $false$.

Output: The number of solutions, or the updated belief, $v(X_1)$.

```
   if caching == true then                                         // Initialize cache tables
1      Initialize cache tables with entries of "−1"
2  v(X_1) ← 0; OPEN ← {X_1}                                        // Initialize the stack OPEN
3  while OPEN ≠ Φ do
4      n ← top(OPEN); remove n from OPEN
5      if caching == true and n is OR, labeled X_i and Cache(val(path(n))[pa_{X_i}]) ≠ −1 then   // In
       cache
6          v(n) ← Cache(val(path(n))[pa_{X_i}])                    // Retrieve value
7          successors(n) ← φ                                       // No need to expand below
8      else                                                        // EXPAND
9          if n is an OR node labeled X_i then                     // OR-expand
10             successors(n) ← {⟨X_i, x_i⟩ | ⟨X_i, x_i⟩ is consistent with path(n) }
11             v(⟨X_i, x_i⟩) ← 1,   for all ⟨X_i, x_i⟩ ∈ successors(n)
12             v(⟨X_i, x_i⟩) ← ∏_{f ∈ B_T(X_i)} f(val(path(n))[pa_{X_i}]),  for all ⟨X_i, x_i⟩ ∈ successors(n)
               // AO-BU
13         if n is an AND node labeled ⟨X_i, x_i⟩ then             // AND-expand
14             successors(n) ← children_T(X_i)
15             v(X_i) ← 0 for all X_i ∈ successors(n)
16         Add successors(n) to top of OPEN
17     while successors(n) == Φ do                                 // PROPAGATE
18         if n is an OR node labeled X_i then
19             if X_i == X_1 then                                  // Search is complete
20                 return v(n)
21             if caching == true then
22                 Cache(val(path(n))[pa_{X_i}]) ← v(n)            // Save in cache
23             v(p) ← v(p) * v(c)
24             if v(p) == 0 then                                   // Check if p is dead-end
25                 remove successors(p) from OPEN
26                 successors(p) ← Φ
27         if n is an AND node labeled ⟨X_i, x_i⟩ then
28             let p be the parent of n
29             v(p) ← v(p) + v(n);
30         remove n from successors(p)
31         n ← p
```

6.5.1 COMPLEXITY

From Theorem 6.33 we can clearly conclude the following.

Theorem 6.42 *For any reasoning problem, algorithm* AOT *runs in linear space and in $O(nk^h)$ time, when h is the height of the guiding pseudo tree and k is the maximum domain size. If the primal* *graph has a tree decomposition with treewidth* w, *then there exists a pseudo tree* $\mathcal{T}$ *for which AOT is* $O(nk^{w \cdot \log n})$. *(Exercise: provide a proof).*

Obviously, for constraint satisfaction the algorithm would terminate early with a first solution, andwould potentially be much faster than for the rest of the queries. Based on Theorem 6.29 we can derive a complexity bound when searching the AND/OR context-minimal search graph.

Theorem 6.43 *For any reasoning problem, the complexity of algorithm* AOG *(i.e., algorithm 2 for the flag caching=true) is time and space* $O(nk^w)$ *where* w *is the induced width of the guiding pseudo tree and* k *is the maximum domain size.*

The space complexity of AOG can often be far tighter than exponential in the treewidth. One reason is related to the space complexity of tree decomposition schemes which, as we know, can operate in space exponential in the size of the cluster separators only, rather than be exponential in the cluster size. We will use the term *dead caches* to address this issue. Intuitively, a node that has only one incoming arc in the search tree will be traversed only once by DFS search, and therefore its value does not need to be remembered, as it will never be used again. Luckily, such nodes can be recognized based only on their context.

Definition 6.44 Dead cache. [Darwiche, 2001a] If X is the parent of Y in pseudo tree $\mathcal{T}$, and $context(X) \subset context(Y)$, then $context(Y)$ is a *dead cache.*

We know that a pseudo-tree is also a bucket-tree. That is, given a pseudo-tree we can generate a bucket-tree by associating a cluster for each variable X_i and its parents pa_{X_i} in the induced-graph. Following the pseudo-tree structure, some of the clusters may not be maximal, and these are precisely the ones that correspond to dead caches. The parents pa_{X_i} that are not dead caches are those separators between maximal clusters in the bucket tree associated with the pseudo-tree.

Example 6.45 Consider the graphical models and the pseudo tree in Figure 6.12a. The context in the left branch $(C, CK, CKL, CKLN)$ are all dead-caches. The only one which is not a dead cache is CKO, the context of P. As you can see, there are converging arcs into P only along this branch. Indeed if we describe the clusters of the corresponding bucket-tree. we would have just two maximal clusters: $CKLNO$ and $PCKO$ whose separator is CKO, which is the context of P. (Exercise: Determine the dead caches for the pseudo-tree in Figure 6.12d).

We can conclude the following.

Proposition 6.46 *If dead caches are not recorded, the space complexity of AOG can be reduced to being exponential in the separator's size only, while still being time exponential in the induced-width. (Prove as an exercise.)*

Finding a single solution. We can easily modify the algorithm to find a single solution. The main difference is that the 0/1 v values of internal nodes are propagated using *Boolean* summation and product instead of regular operators. If there is a solution, the algorithm terminates as soon as the value of the root node is updated to 1. The solution subtree can be generated by following the pointers of the latest solution subtree. This, of course, is a very naive way of computing a single consistent solution as would be discussed in the context of mixed networks in Section 6.6.

Finding posterior marginals. To find posterior marginal of the root variable, we only need to keep the computation at the root of the search graph and normalize the result. However, if we want to find the belief (i.e., posterior marginal) for each variable we would need to make a more significant adaptation of the search scheme.

Optimization tasks. General AND/OR algorithms for evaluating the value of a root node for any reasoning problem using tree or graph AND/OR search spaces are identical to the above algorithms when product is replaced by the appropriate combination operator (i.e., product or summation) and marginalization by summation is replaced by the appropriate marginalization operator. For optimization (e.g., mpe) all we need is to change line 30 of the algorithm from summation to maximization. Namely, we should have $v(p) \leftarrow max\{v(p), v(n)\}$. Clearly, this will yield a base-line scheme that can be advanced significantly using heuristic search ideas. To compute marginal map query we will use marginalization by sum of max based on the variable identity to which the marginalization operator is applied.

Depth-first vs Best-first searches This could be the right place to make a clear distinction between searching the AND/OR space depth-first of best-first for optimization tasks. Best-first cannot exploit dead-caches, but must cache all nodes in the explicated graph. For this reason DFS can have far better memory utilization even when both scheme search an AD/OR graph.

6.6 AND/OR SEARCH ALGORITHMS FOR MIXED NETWORKS

We will consider now AND/OR search schemes in the context of mixed graphical models (see definition 2.25 in Section 2.6). To refresh, the mixed network is defined by a pair of a Bayesian network and a constraint network. This pair expresses a probability distribution over all the variables which is conditioned on the requirement that all the assignments having nonzero probability satisfy all the constraints. The constraint network may be specified explicitly as such, or can be extracted from the probabilistic network as those partial tuples whose probability is zero (see Definition 2.25 and [Darwiche, 2009] chapter 13).

All advanced constraint processing algorithms [Dechter, 2003], either incorporating no-good learning and constraint propagation during search, or using variable elimination algorithms such as *adaptive-consistency* and *directional resolution*, can be applied to AND/OR search for mixed networks. In this section we will touch on these methods briefly because they need a considerable

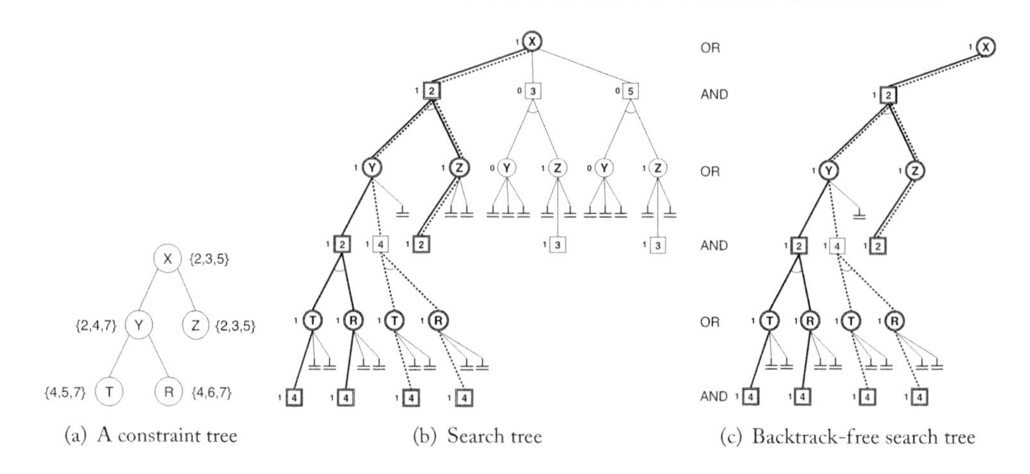

(a) A constraint tree (b) Search tree (c) Backtrack-free search tree

Figure 6.14: AND/OR search tree and backtrack-free tree.

space to be treated adequately. Our aim is to point to the main principles of constraint processing that can have impact in the context of AND/OR search and refer the reader to the literature for full details (see [Dechter, 2003]).

Overall, the virtue of having the mixed network view is that the constraint portion can be processed by a wide range of constraint processing techniques, both statically before search, or dynamically during search [Dechter, 2003]. Recall the concept of *backtrack-free* (see Chapter 3).

Definition 6.47 Backtrack-free AND/OR search tree. Given graphical model $\mathcal{M}$ and given an AND/OR search tree $S_{\mathcal{T}}(\mathcal{M})$, the *backtrack-free AND/OR search tree* of $\mathcal{M}$ w.r.t. $\mathcal{T}$ is obtained by pruning from $S_{\mathcal{T}}(\mathcal{M})$ all inconsistent subtrees, namely all nodes that root no consistent partial solution (have a value 0).

Example 6.48 Consider 5 variables X, Y, Z, T, R over domains $\{2, 3, 5\}$, where the constraints are: X divides Y and Z, and Y divides T and R. The constraint graph and the AND/OR search tree relative to the guiding DFS tree rooted at X, are given in Figure 6.14a, b. In 6.14b we present the $S_{\mathcal{T}}(\mathcal{R})$ search space whose nodes' consistency status are already evaluated as having value "1" if consistent and "0" otherwise. We also highlight two solutions subtrees; one depicted by solid lines and one by dotted lines. Part (c) presents the backtrack-free tree where all nodes that do not root a consistent solution are pruned.

If we traverse the backtrack-free AND/OR search tree we can find a solution subtree without encountering any dead-ends. Some constraint networks specifications yield a backtrack-free search space. Others can be made backtrack-free by massaging their representation using *constraint propagation* algorithms. In particular, the variable-elimination algorithms *adaptive-consistency* described in Chapter 3, and directional resolution, compiles a constraint specification (resp., a Boolean CNF formula) into one that has a backtrack-free search space along some orderings. We remind now the definition of the directional extension (see also Chapter 3).

Definition 6.49 Directional extension [Dechter and Pearl, 1987; Rish and Dechter, 2000]. Let $\mathcal{R} = \langle \mathbf{X}, \mathbf{D}, \mathbf{C}, \bowtie \rangle$ be a constraint problem and let d be a DFS ordering of a pseudo tree, then the directional extension $E_d(\mathcal{R})$ denotes the constraint network (resp., the CNF formula) compiled by adaptive-consistency (resp., directional resolution) in reversed order of d.

Example 6.50 In Example 6.48, if we apply adaptive-consistency in reverse order of $d = (X, Y, T, R, Z)$, the algorithm will remove the values $\{3, 5\}$ from the domains of both X and Z yielding a tighter directional extension network $\mathcal{R}'$. As noted before, the AND/OR search tree of $\mathcal{R}'$ in Figure 6.14c is backtrack-free.

Proposition 6.51 *Given a constraint network $\mathcal{R} = \langle \mathbf{X}, \mathbf{D}, \mathbf{C}, \bowtie \rangle$, and a pseudo-tree $\mathcal{T}$, the AND/OR search tree of the graphical model compiled into a directional extension $E_d(\mathcal{R})$ when d is a DFS ordering of $\mathcal{T}$, coincides with the backtrack-free AND/OR search tree of $\mathcal{R}$ based on $\mathcal{T}$. (See Appendix for proof.)*

Proposition 6.51 emphasizes the significance of no-good learning for deciding inconsistency or for finding a single solution. These techniques are known as clause learning in SAT solvers [Jr. and Schrag, 1997] and are currently used in most advanced solvers [Marques-Silva and Sakalla, 1999]. Namely, when we apply no-good learning we explore a pruned search space whose many inconsistent subtrees are removed, but we do so more gradually, during search, then when applying the full variable-elimination compilation, before search.

We will now give more details on applying constraint techniques while searching the AND/OR search space for processing queries over mixed networks. The mixed network can be processed by tightening the constraint network only. Namely we can process the deterministic information separately (e.g., by enforcing some consistency level [Dechter and Mateescu, 2007b]).

6.6.1 AND-OR-CPE ALGORITHM

Algorithm AND-OR-CPE for the constraint probabilistic evaluation query (CPE) (Definition 2.27) is given in Algorithm 3. The input is a mixed network, a pseudo tree $\mathcal{T}$ of the mixed graph and the context of each variable. The output is the probability that a random tuple generated from the belief network distribution is consistent (satisfies the constraint portion). As common with other queries, AND-OR-CPE traverses the AND/OR search tree or graph corresponding to $\mathcal{T}$ in a DFS manner and each node maintains a value v which accumulates the computation in its own

Algorithm 3: AND-OR-CPE.

Input: A mixed network $\mathcal{B} = \langle \mathbf{X}, \mathbf{D}, \mathbf{P}_G, \prod \rangle$ that expresses $P_{\mathcal{B}}$ and a constraint network $\mathcal{R} = \langle \mathbf{X}, \mathbf{D}, \mathbf{C}, \bowtie \rangle$; a pseudo tree $\mathcal{T}$ of the moral mixed graph, rooted at X_1; parents pa_i (OR-context) for every variable X_i; caching set to $true$ or $false$.

Output: The probability $P(\bar{x} \in \rho(\mathcal{R}))$ that a tuple satisfies the constraint query.

if caching $== true$ **then** // Initialize cache tables
1. Initialize cache tables with entries of "-1"
2. $v(X_1) \leftarrow 0$; OPEN $\leftarrow \{X_1\}$ // Initialize the stack OPEN
3. **while** OPEN $\neq \Phi$ **do**
4. n $\leftarrow top$(OPEN); remove n from OPEN
5. **if** caching $== true$ **and** n *is OR, labeled* X_i **and** $Cache(val(path(n))[pa_{X_i}]) \neq -1$ **then** // If in cache
6. v(n) $\leftarrow Cache(val(path(n))[pa_{X_i}])$ // Retrieve value
7. $successors$(n) $\leftarrow \Phi$ // No need to expand below
8. **else** // **Expand search (forward)**
9. **if** n *is an OR node labeled* X_i **then** // OR-expand
10. $successors$(n) $\leftarrow$ ConstraintPropagation($\langle \mathbf{X}, \mathbf{D}, \mathbf{C} \rangle, val(path(n))$) // **CONSTRAINT PROPAGATION**
11. $v(\langle X_i, x_i \rangle) \leftarrow \prod_{f \in B_{\mathcal{T}}(X_i)} f(val(path(n))[pa_{X_i}])$, for all $\langle X_i, x_i \rangle \in successors$(n)
12. **if** n *is an AND node labeled* $\langle X_i, x_i \rangle$ **then** // AND-expand
13. $successors$(n) $\leftarrow children_{\mathcal{T}}(X_i)$
14. $v(X_i) \leftarrow 0$ for all $X_i \in successors$(n)
15. Add $successors$(n) to top of OPEN
16. **while** $successors$(n) $== \Phi$ **do** // **Update values (backtrack)**
17. **if** n *is an OR node labeled* X_i **then**
18. **if** $X_i == X_1$ **then** // Search is complete
19. **return** v(n)
20. **if** caching $== true$ **then**
21. $Cache(val(path(n))[pa_i]) \leftarrow v$(n) // Save in cache
22. let p be the parent of n
23. v(p) $\leftarrow v$(p) $* v$(n)
24. **if** v(p) $== 0$ **then** // Check if p is dead-end
25. remove $successors$(p) from OPEN
26. $successors$(p) $\leftarrow \Phi$
27. **if** n *is an AND node labeled* $\langle X_i, x_i \rangle$ **then**
28. let p be the parent of n
29. v(p) $\leftarrow v$(p) $+ v$(n);
30. remove n from $successors$(p)
31. n $\leftarrow$ p

subtree. As before we have a recursive computation. OR nodes accumulate the summation of the product between each child's value and its OR-to-AND weight, while AND nodes accumulate the product of their children's values. The context-based caching is done using table data structures as described earlier.

Procedure `ConstraintPropagation`($\mathcal{R}$, $\mathbf{x}_1^i$).

Input: A constraint network $\mathcal{R} = \langle \mathbf{X}, \mathbf{D}, \mathbf{C} \rangle$; a partial assignment path $\bar{x}_i$ to variable X_i.

Output: reduced domain D_i of X_i; reduced domains of future variables; newly inferred constraints.
This is a generic procedure that performs the desired level of constraint propagation, for example forward checking, unit propagation, arc consistency over the constraint network $\mathcal{R}$ and conditioned on $\mathbf{x}_1^i$.

return *reduced domain of* X_i

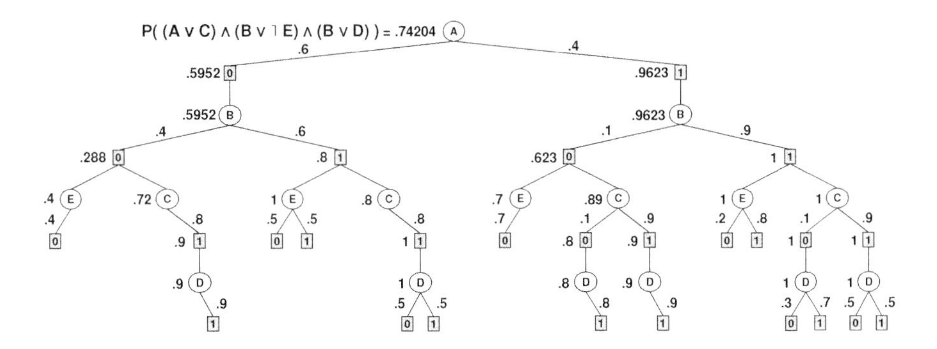

Figure 6.15: Mixed network defined by the query $\varphi = (A \vee C) \wedge (B \vee \neg E) \wedge (B \vee D)$.

Example 6.52 We refer back to the example in Figure 6.3. Consider a constraint network that is defined by the CNF formula $\varphi = (A \vee C) \wedge (B \vee \neg E) \wedge (B \vee D)$. The trace of algorithm AND-OR-CPE without caching is given in Figure 6.15. Notice that the clause $(A \vee C)$ is not satisfied if $A = 0$ and $C = 0$, therefore the paths that contain this assignment cannot be part of a solution of the mixed network. The value of each node is shown to its left (the leaf nodes assume a dummy value of 1, not shown in the figure). The value of the root node is the probability of φ. Notice how similar is Figure 6.15 to Figure 6.3. Indeed, in 6.3 we seek the probability of evidence which can be modeled as the CNF formula having unit clauses $D \wedge \neg E$.

6.6.2 CONSTRAINT PROPAGATION IN AND-OR-CPE

We next discuss the use of constraint propagation during search. This methods are used in any constraint or SAT/CSP (see Chapters 5 and 6 in [Dechter, 2003]). In general, constraint propagation helps to discover what variable and what value to not instantiate in order to avoid dead-ends as much as possible. This is done with a bounded level of computation. The incorporation of these methods on top of AND/OR search for computation of the value of the root is straightforward. For illustration, we will only consider static variable ordering based on a pseudo tree, and will focus on the impact of constraint propagation on domain-value order of assignments to the variables.

In algorithm AND/OR-CPE, line 10 contains a call to the generic `ConstraintPropagation` procedure consulting only the constraint subnetwork $\mathcal{R}$, conditioned on the current partial assignment. The constraint propagation is relative to the current set of constraints, the given path that defines the current partial assignment, and the newly inferred constraints, if any, that were learned during search. `ConstraintPropagation` which requires polynomial time, may discover that some domain value-assignments to the variables cannot be extended to a full solution. These assignments are marked as dead-ends and removed from the current domain of the variable. All the remaining domain values remain feasible and are returned by the procedure as possible candidates to extend the search frontier. Clearly, not all those assignments are guaranteed to lead to a solution.

We therefore have the freedom to employ any procedure for checking the consistency of the constraints of the mixed network. The simplest case is when no constraint propagation is used and only the initial constraints of $\mathcal{R}$ are checked for consistency. We denote this algorithm by AO-C.

We also consider two forms of constraint propagation on top of AO-C. The first algorithm AO-FC, is based on *forward checking*, which is one of the weakest forms of propagation. It propagates the effect of a domain-value assignment to each future uninstantiated variable separately, and checks consistency against the constraints whose scope would become fully instantiated by just one such future variable.

The second algorithm, referred to as AO-RFC, performs a variant of *relational forward checking*. Rather than checking only constraints whose scope becomes fully assigned, AO-RFC checks all the existing constraints by looking at their projection on the variables along the current path. If the projection is empty an inconsistency is detected. AO-RFC is computationally more expensive than AO-FC, but yields a more pruned search space.

Example 6.53 Figure 6.16a shows the belief part of a mixed network, and Figure 6.16b the constraint part. All variables have the same domain, {1,2,3,4}, and the constraints express "less than" relations. Figure 6.16c shows the search space of AO-C. Figure 6.16d shows the space traversed by AO-FC. Figure 6.16e shows the space when consistency is enforced with Maintaining Arc Consistency (which enforces full arc-consistency after each new instantiation of a variable).

SAT solvers. One possibility that was explored with success (e.g., [Allen and Darwiche, 2003]) is to delegate the constraint processing to a separate off-the-shelf SAT solver. In this case, for each new variable assignment the constraint portion is packed and fed into the SAT solver. If no solution is reported, then that value is a dead-end. If a solution is found by the SAT solver, then the AND/OR search continues (remember that for some tasks we may have to traverse all the solutions of the graphical model, so the one solution found by the SAT solver does not finish the task). Since, the worst-case complexity of this level of constraint processing, at each node, is exponential in the worst-case, a common alternative is to use *unit propagation*, or *unit resolution*, as a form of bounded resolution (see Chapter 3 and [Rish and Dechter, 2000]).

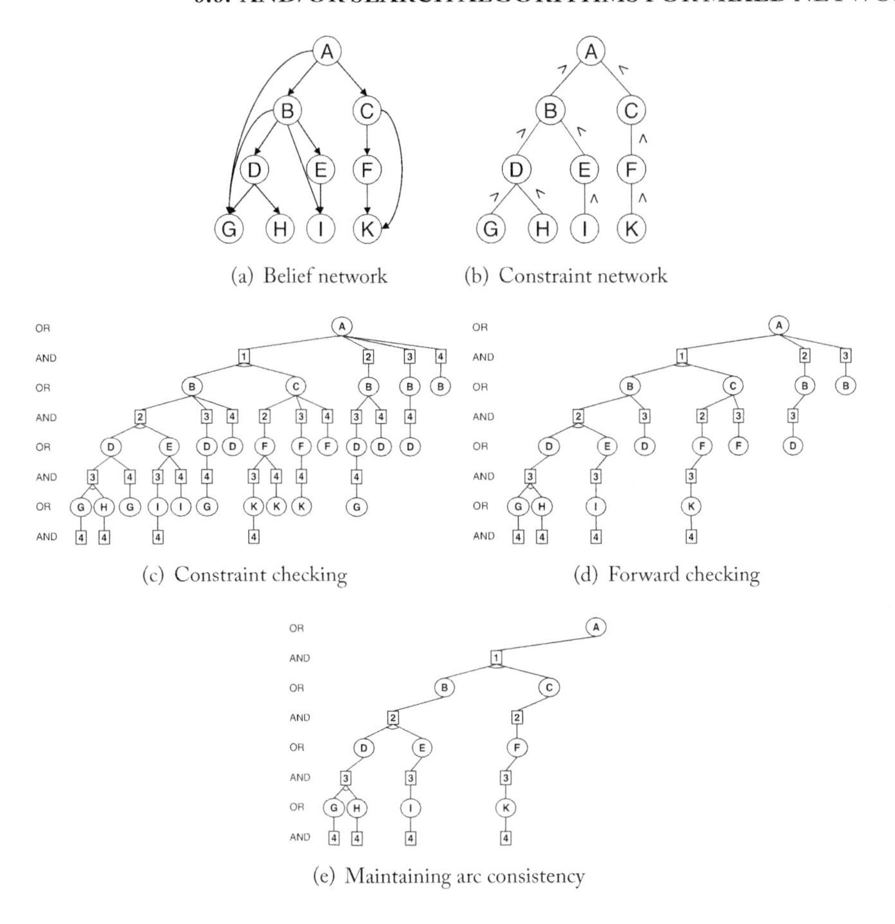

(a) Belief network (b) Constraint network

(c) Constraint checking (d) Forward checking

(e) Maintaining arc consistency

Figure 6.16: Traces of AND-OR-CPE with various levels of constraint propagation.

Such a hybrid use of search and a specialized efficient SAT (or constraint) solver can be very useful, and it underlines further the power that the mixed network representation has in delimiting the constraint portion from the belief network.

6.6.3 GOOD AND NOGOOD LEARNING

When a search algorithm encounters a dead-end, it can use different techniques to identify the ancestor variable assignments that caused the dead-end, which is called a *conflict-set*. It is conceivable that the same assignment of that set of ancestor variables may be encountered in the future, and they would then lead to the same dead-end. Rather than rediscovering it again, if memory allows, it is useful to record the dead-end conflict-set as a new constraint (or clause) over the ancestor variable set that is responsible for it. Recording dead-end conflict-sets is sometimes called nogood learning.

One form of nogood learning is graph-based, and it uses various techniques to identify the ancestor variables that generate the nogood. The information on conflicts is generated from the primal graph information alone. It is easy to show that AND/OR search already implements this information within the context of the nodes. Therefore, if caching is used, just saving the information about the nogoods encountered amounts to nogood learning techniques such as graph-based learning (see [Dechter, 2003]).

If deeper types of nogood learning are desirable, they need to be added on top of the AND/OR search. In such a case, a smaller set than the context of a node may be identified as a culprit assignment, and may help discover future dead-ends much earlier than when context-based caching alone is used. Needless to say, deeper learning is computationally more expensive.

In recent years [Beam, Kautz, Tian Sang, Bacchus and Piassi, 2004; Darwiche, 2001a; Dechter and Mateescu, 2007b], several schemes propose not only the learning of nogoods, but also that of their logical counterparts, the *goods*. Traversing the context minimal AND/OR graph and caching appropriately can be shown to implement both good and nogood graph-based learning.

6.7 SUMMARY AND BIBLIOGRAPHICAL NOTES

Chapters 6 presents search for graphical models using the concept of AND/OR search spaces rather than OR spaces. It introduced the AND/OR search tree, and showed that its size can be bounded exponentially by the depth of its pseudo tree over the graphical model. This implies exponential savings for any linear space algorithms traversing the AND/OR search tree. Specifically, if the graphical model has treewidth w^*, the height of the pseudo tree is $O(w^* \cdot \log n)$.

The AND/OR search tree can be further compacted into a graph by merging identical subtrees. We showed that the size of the minimal AND/OR search graph is exponential in the treewidth, while the size of the minimal OR search graph is exponential in the pathwidth. Since for some graphs the difference between treewidth and pathwidth is substantial, the AND/OR representation implies substantial time and space savings for memory intensive algorithms traversing the AND/OR graph. Searching the AND/OR search *graph* can be implemented by caching during search, while no-good recording is interpreted as pruning portions of the search space independent of it being a tree or a graph, an OR or an AND/OR.

The chapter is based on the work by Dechter and Mateescu [Dechter and Mateescu, 2007b]. The AND/OR search space is inspired by search advances introduced sporadically in the past three decades for constraint satisfaction and more recently for probabilistic inference and for optimization tasks. Specifically, it resembles pseudo tree rearrangement [Freuder and Quinn, 1987; Freuder, 1985], briefly introduced more than two decades ago, which was adapted subsequently for distributed constraint satisfaction [Collin, Dechter and Katz, 1991, 1999] and more recently in [Modi *et al.*, 2005], and was also shown to be related to graph-based backjumping [Dechter, 1992]. This work was extended in [Bayardo and Miranker, 1996] and more recently applied to optimization tasks [Larrosa and Sanchez, 2002]. Another version that can be viewed as exploring the AND/OR graphs was presented recently for constraint satisfaction [Terrioux and Jegou, 2003b] and for optimization [Terrioux and Jegou, 2003a]. Similar principles were introduced for probabilistic inference in algorithm Recursive Conditioning [Darwiche, 2001a] as well as in Value Elimination [F. Bacchus and Piassi, 2003a,b] and currently provide the backbones of the most advanced SAT solvers [Beam, Kautz, Tian Sang, Bacchus and Piassi, 2004].

It is known that exploring the search space in a dynamic variable ordering is highly beneficial. AND/OR search trees for graphical models can also be modified to allow dynamic variable ordering. This require a careful balancing act of the computational overhead that normally accompanies dynamic search schemes. For further information see [F. Bacchus and Piassi, 2003b; Marinescu and Dechter, 2009a].

6.8 APPENDIX: PROOFS

Proof of Theorem 6.13

(1) By definition, all the arcs of $S_{\mathcal{T}}(\mathcal{R})$ are consistent. Therefore, any solution tree of $S_{\mathcal{T}}(\mathcal{R})$ denotes a solution for $\mathcal{R}$ whose assignments are all the labels of the AND nodes in the solution tree. Also, by definition of the AND/OR tree, every solution of $\mathcal{R}$ must corresponds to a solution subtree in $S_{\mathcal{T}}(\mathcal{R})$. (2) By construction, the set of arcs in every solution tree have weights such that each function of F contribute to one and only one weight via the combination operator. Since the total weight of the tree is derived by combination, it yields the cost of a solution. □.

Proof of Theorem 6.14

Let p be an arbitrary directed path in the pseudo tree $\mathcal{T}$ that starts with the root and ends with a leaf. This path induces an OR search subtree which is included in the AND/OR search tree $S_{\mathcal{T}}$, and its size is $O(k^h)$ when h bounds the path length. The pseudo tree $\mathcal{T}$ is covered by l such directed paths, whose lengths are bounded by h. The union of their individual search trees covers the whole AND/OR search tree $S_{\mathcal{T}}$, where every distinct full path in the AND/OR tree appears exactly once, and therefore, the size of the AND/OR search tree is bounded by $O(l \cdot k^h)$. Since $l \leq n$ and $l \leq b^m$, it concludes the proof. The bounds are tight because they are realizable for graphical models whose all full assignments are consistent. □.

Proof of Theorem 6.19

(1) All we need to show is that the *merge* operator is not dependant on the order of applying the operator. Mergeable nodes can only appear at the same level in the AND/OR graph. Looking at the initial AND/OR graph, before the merge operator is applied, we can identify all the mergeable nodes per level. We prove the proposition by showing that if two nodes are initially mergeable, then they must end up merged after the operator is applied exhaustively to the graph. This can be shown by induction over the level where the nodes appear.

Base case: If the two nodes appear at the leaf level (level 0), then it is obvious that the exhaustive merge has to merge them at some point.

Inductive step: Suppose our claim is true for nodes up to level k and two nodes n_1 and n_2 at

level $k+1$ are initially identified as mergeable. This implies that, initially, their corresponding children are identified as mergeable. These children are at level k, so it follows from the inductive hypothesis that the exhaustive merge has to merge the corresponding children. This in fact implies that nodes n_1 and n_2 will root the same subgraph when the exhaustive merge ends, so they have to end up merged. Since the graph only becomes smaller by merging, based on the above the process, merging has to stop at a fix point.
(2) Analogous to (1).
(3) If the nodes can be merged, it follows that the subgraphs are identical, which implies that they define the same conditioned subproblems, and therefore the nodes can also be unified. □

Proof of Proposition 6.51.
Note that if $\mathcal{T}$ is a pseudo tree of $\mathcal{R}$ and if d is a DFS ordering of $\mathcal{T}$, then $\mathcal{T}$ is also a pseudo tree of $E_d(\mathcal{R})$ and therefore $S_{\mathcal{T}}(E_d(\mathcal{R}))$ is a faithful representation of $E_d(\mathcal{R})$. $E_d(\mathcal{R})$ is equivalent to $\mathcal{R}$, therefore $S_{\mathcal{T}}(E_d(\mathcal{R}))$ is a supergraph of $BF_{\mathcal{T}}(\mathcal{R})$. We only need to show that $S_{\mathcal{T}}(E_d(\mathcal{R}))$ does not contain any dead-ends, in other words any consistent partial assignment must be extendable to a solution of $\mathcal{R}$, This however is obvious, because Adaptive consistency makes $E_d(\mathcal{R})$ strongly directional $w^*(d)$ consistent, where $w^*(d)$ is the induced width of R along ordering d [Dechter and Pearl, 1987], a notion that is synonym with backtrack-freeness. □

CHAPTER 7

Combining Search and Inference: Trading Space for Time

We introduced inference and search algorithm separately, because they seem to have useful but complementary properties. We saw that if we apply search over an AND/OR space and if, in addition, we use context-based caching, search has the same worst-case complexity as inference, both are time and space exponentially bounded by the treewidth. However, when the treewidth is too high, the needed memory is not available and schemes that are more flexible in trading space for time are needed. This is one of the main virtues of search compared with inference. Search can be done in linear memory and accommodates space for time tradeoff quite flexibly. In fact, we have already observed that the class of AND/OR search $AO(i)$ of searching the AND/OR context minimal search space and caching tables based on bounded (by i) context only, can be trade time and space effectively. In this chapter we discuss several additional approaches trading space for time that are based on combining search with inference. We demonstrate the principles of such hybrids in the context of OR tree search, first, and then extend it to AND/OR search.

7.1 THE CUTSET-CONDITIONING SCHEME

The algorithms presented in this section exploit the fact that variable instantiation changes the effective connectivity of the primal graph.

7.1.1 CUTSET-CONDITIONING FOR CONSTRAINTS

Consider a constraint problem whose primal graph is given in Figure 7.1a. For this problem, instantiating X_2 to some value, say a, renders the choices of values to X_1 and X_5 independent, as if the pathway $X_1 - X_2 - X_5$ were blocked at X_2. Similarly, this instantiation blocks dependency in the pathway $X_1 - X_2 - X_4$, leaving only one path between any two variables. In other words, given that X_2 was assigned a specific value, the "effective" constraint graph for the rest of the variables is shown in Figure 7.1b. Here, the instantiated variable X_2 and all its incident arcs are first deleted from the graph, and X_2 subsequently is duplicated for each of its neighbors. The constraint problem having the graph shown in Figure 7.1a when $X_2 = a$ is identical to the constraint problem having the graph in Figure 7.1b with the same assignment $X_2 = a$. We already

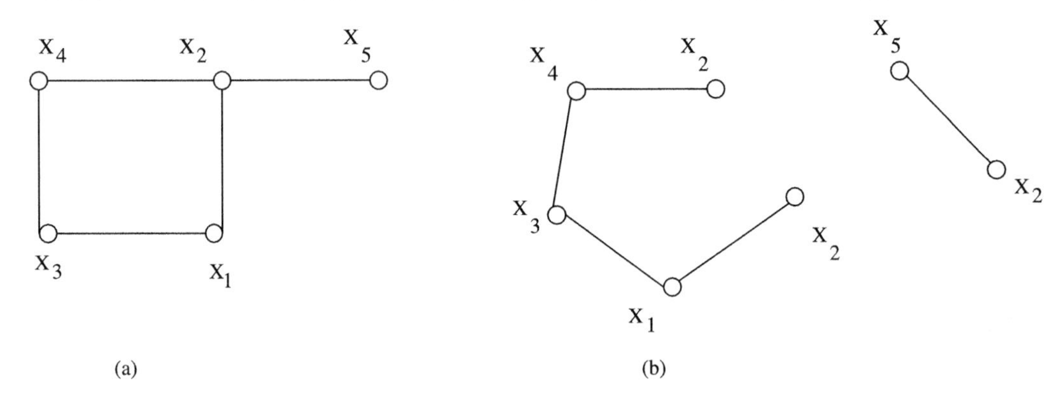

Figure 7.1: An instantiated variable cuts its own cycles.

saw, in Chapter 6 that this idea leads to problem decomposition and to the notion of AND/OR graph. Here, however, we look at a more intermediate phenomena where variable conditioning makes the remaining problem more tractable even if it does not decompose.

In general, when the group of instantiated variables constitutes a cycle-cutset; a set of nodes that, once removed, would render the constraint graph cycle-free, as shown in Figure 7.1b, and can be solved by *tree-solving* algorithm. The tree solving algorithm can be belief propagation, or its constraint version of arc-consistency. In most practical cases it would take more than a single variable to cut all the cycles in the graph. Thus, a general way of solving a problem whose constraint graph contains cycles is to identify a subset of variables that cut all cycles in the graph, find a consistent instantiation of the variables in the cycle-cutset, and then solve the remaining problem by the inference *tree algorithm*. If a solution to this restricted problem (conditioned on the cycle-cutset values) is found, then a consistent solution to the entire problem is at hand. If not, another instantiation of the cycle-cutset variables should be considered until a solution is found. If we seek to enumerate all solutions or to count them, we would have to enumerate over all the assignments to the cutset variables.

Example 7.1 If the task is to solve a constraint problem whose constraint graph is presented in Figure 7.1a (assume X_2 has two values $\{a, b\}$ in its domain), first $X_2 = a$ must be assumed, and the remaining tree problem relative to this instantiation, is solved. If no solution is found, it is assumed that $X_2 = b$ and another attempt is made.

The number of times the tree-solving algorithm needs to be invoked is bounded by the number of partial solutions to the cycle-cutset variables. A small cycle-cutset is therefore desirable. Finding the minimal cycle-cutset, is computationally hard [Garey and Johnson, 1979] however, so it will be more practical to settle for heuristic compromises. The problem, which is also known as the *feedback set problem* was investigated extensively and approximation and heuristic approaches were presented (e.g., [Bar-Yehuda *et al.*, 1998; Becker *et al.*, 2000]). One simple approach is to incorporate the cutset scheme within depth-first backtracking search. Because *DFS backtracking* works by progressively instantiating sets of variables, we only need to keep track of the connectivity status of the primal graph. As soon as the set of instantiated variables constitutes a cycle-cutset, the search algorithm can switch to the tree-solving inference algorithm on the restricted conditioned problem, i.e., either finding a consistent extension for the remaining variables (thus finding a solution to the entire problem) or concluding that no such extension exists (in which case backtracking takes place and another instantiation tried).

Example 7.2 Assume that DFS-backtracking search instantiates the variables of the CSP represented in Figure 7.2a in the order C, B, A, E, D, F (Figure 7.2b). Backtracking will instantiate variables C, B, and A, and then, realizing that these variables cut all cycles, will invoke a tree-solving algorithm on the rest of the problem. That is, the tree problem in Figure 7.2c, with variables C, B, and A assigned, should then be attempted. If no solution is found, control returns to backtracking which will go back to variable A and assign it a new domain value.

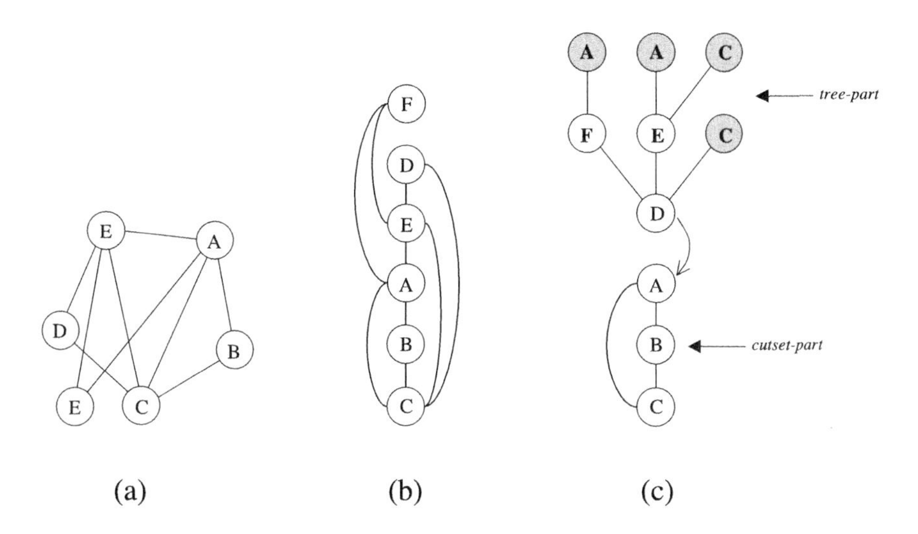

Figure 7.2: (a) A constraint graph, (b) its ordered graph, and (c) the constraint graph of the cutset variable and the conditioned variable, where the assigned variables are darkened.

This idea, often referred to as *cutset-conditioning*, generalizes to all graphical models. As observed in Chapter 4 and in particular in Section 4.1.3, when variable are assigned a value, the connectivity of the graph reduces, thus yielding saving in computation. This yielded the notion of *conditional induced-width* (see 4.7.) which controls the impact of observed variables by bounding the complexity of the respective algorithms more tightly. Rather than insisting on conditioning on a subset of variables that cuts all cycles and yields subproblems having induced-width 1, we can allow cutsets that create subproblems whose induced-width is higher than 1 but still bounded. This suggests a framework of hybrid algorithms parameterized by an integer q which bounds the induced-width of the conditioned subproblems solved by inference.

Definition 7.3 **q-cutset, minimal.** Given a graph G, a subset of nodes is called a *q-cutset* for an integer q iff when removed, the resulting graph has an induced-width less than or equal to q. A minimal *q-cutset* of a graph has a smallest size among all q-cutsets of the graph. A cycle-cutset is a 1-cutset of a graph.

Finding a minimal q-cutset is clearly a hard task [A. Becker and Geiger, 1999; Bar-Yehuda *et al.*, 1998; Becker *et al.*, 2000; Bidyuk and Dechter, 2004]. However, like in the special case of a cycle-cutset we can settle for a non-minimal q-cutset relative to a given variable ordering. Namely, given an ordering, we can seek an initial prefix set of the ordering that is a q-cutset of the problem. Then a DFS search algorithm can traverse the search space over the q-cutset variables and for each of its consistent assignment solve the rest of the problem by an inference algorithm such as ADPATIVE-CONSISTENCY if it is a constraint problem or by BUCKET-ELIMINATION or CLUSTER-TREE ELIMINATION, in the general case.

Example 7.4 Consider as another example the contsaint graph of a graph coloring problem given in Figure 7.3a. The search space over a 2-cutset, and the induced-graph of the conditioned instances are depicted in 7.3b.

7.1.2 GENERAL CUTSET-CONDITIONING

We will distinguish two schemes: the *sequential* variable-elimination with conditioning and the *alternating* one. Algorithm variable-elimination and conditioning $VEC(q)$ is described in Figure 7.4. The algorithm is presented in the context of finding a single solution of a constraint networks but it is immediately extendable to any query and to every graphical model. It can apply depth-first search over the q-cutset and then at each leaf of the cutset variable applies bucket-elimination on the remaining variables, thus solving conditioned subproblems. In particular, the constraint problem $\mathcal{R} = \langle \mathbf{X}, \mathbf{D}, \mathbf{C}, \bowtie \rangle$ conditioned on an assignment $\mathbf{Y} = \mathbf{y}$, denoted $\mathcal{R}_{\mathbf{y}}$ is $\mathcal{R}$ augmented with the unary constraints dictated by the assignment $\mathbf{y}$. In the worst case, all possible assignments to the q-cutset variables need to be enumerated. If c is the q-cutset size, in the worst-case k^c is the number of conditioned subproblems having induced-width bounded by q that should be solved, each requiring $O(nk^{q+1})$ steps. Clearly, then

Theorem 7.5 **Properties of cutset-decomposition.** *Given a constraint network* $\mathcal{R} = \langle \mathbf{X}, \mathbf{D}, \mathbf{C}, \bowtie \rangle$ *having n variables and a domain size bounded by k, algorithm* VEC(q) *is*

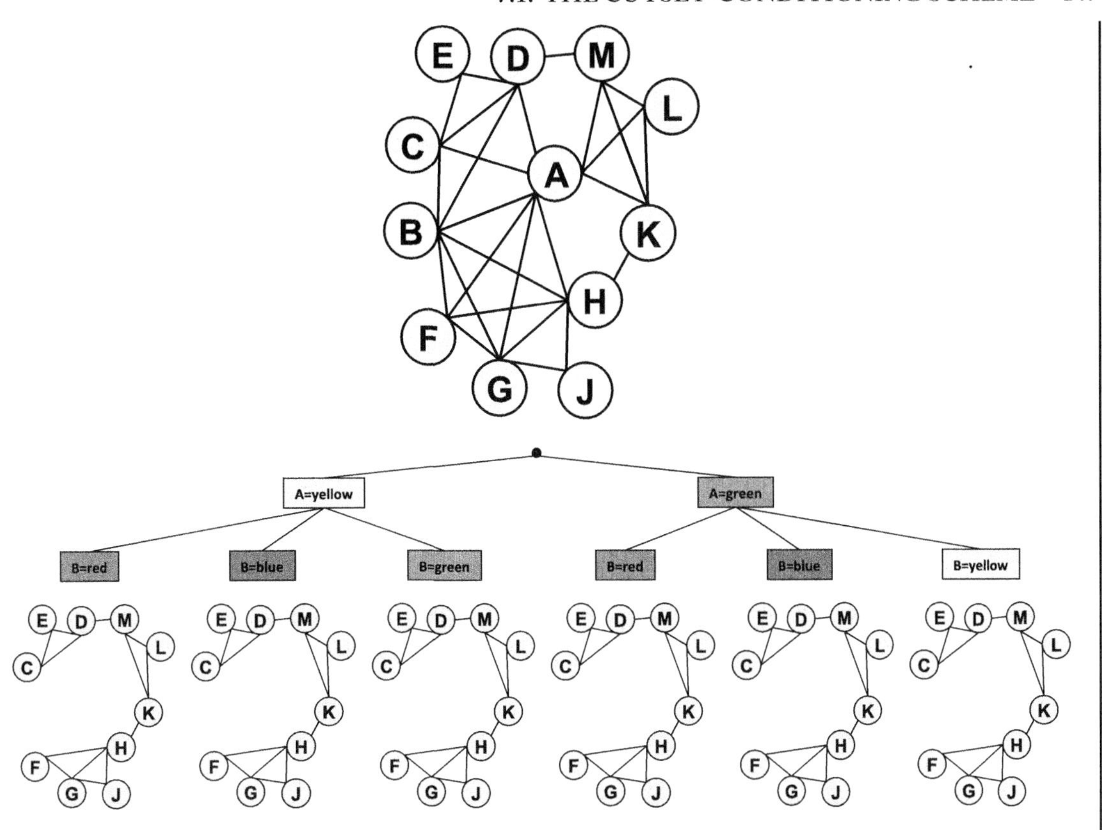

Figure 7.3: (a) A graph coloring problem and (b) its 3-cutset conditioned search space.

sound and complete and it has time and space complexity of $O(n \cdot k^{c+q+1})$ *and* $O(k^q)$*, respectively, where c is the q-cutset size . (Prove as an exercise).* □

The special case of $q = 1$ yields the cycle-cutset algorithm whose time complexity is $O(nk^{c+2})$ which operates in linear space. We see that the integer q can control the balance between search and inference (e.g., variable-elimination), controlling the tradeoff between time and space.

7.1.3 ALTERNATING CONDITIONING AND ELIMINATION

An alternative use of the q-cutset principle is *to alternate* between conditioning-search and variable-elimination. Given a variable ordering we can apply BE to the variables, one by one, as long as the induced-width of the eliminated variables does not exceed q. However, if a variable has induced-width higher than q, the variable will be conditioned upon creating, k subproblems, one for each assigned value on which the alternating algorithm resumes. We call this *Alternating*

Algorithm VEC(q)
Input: A graphical constraint model $\mathcal{R} = \langle \mathbf{X}, \mathbf{D}, \mathbf{C}, \bowtie \rangle$. an ordering d that starts with $\mathbf{Y}$ such that $\mathbf{Y} \subseteq \mathbf{X}$ is a q-cutset, $Z = X - Y$
Output: A consistent assignment, if there is one.

- **while** $\mathbf{y} \leftarrow$ apply backtracking search generating the next partial solution of $\mathbf{Y} = \mathbf{y}$, **do**
  1. $\mathbf{z} \leftarrow$ BE-CSP($\mathcal{R}_{\mathbf{Y}=\mathbf{y}}$).
  2. **if** $\mathbf{z}$ is not *false*, return solution $(\mathbf{y}, \mathbf{z})$.
- **endwhile.**
- **return:** the problem has no solutions.

Figure 7.4: Algorithm variable-elimination with conditioning *VEC(q)*.

Variable-Elimination and Conditioning and denote it as ALT-VEC(q). Clearly, the conditioning set uncovered via ALT-VEC(q) is a q-cutset and therefore can be used as the q-cutset within VEC. (Exercise: Prove that the conditioned set in *ALT-VEC(q)* is a q-cutset.)

Both VEC(q) and ALT-VEC(q) will benefit from the optimization task of finding a minimal q-cutset of a graph which is obviously hard, but some greedy heuristic algorithms were investigated empirically. For $q > 1$ the task was addressed sporadically [Bidyuk and Dechter, 2004; Fishelson and Geiger, 2003]. As in the case of the cycle-cutset (i.e., $q = 1$), we can always use a brute-force approach that fits algorithm ALT-VEC as follows. Given an ordering $d = x_1, ..., x_n$ of G, process the nodes from last to first. When node X is processed, and if its induced-width is greater than q, it is added to the q-cutset and is then removed from the remaining graph. Otherwise, the variable's earlier neighbors are connected, and only then it is removed from the graph. Note that verifying that a given subset of nodes in a graph is a q-cutset of the graph can be accomplished in polynomial time (linear in the number of nodes), by deleting the candidate cutset nodes from the graph and verifying that the remaining graph has an induced width bounded by q. The latter task, however, is exponential in q and therefore quite costly if q is not small.

Example 7.6 As a simple illustration consider Figure 7.5 (left). Applying ALT-VEC(2) will eliminate the variable B first and A next, yielding the graph in the middle and then the graph on the right whose induced-width is higher than 2. We now condition on variable C which generates several subproblems having induced-width of 2 each (see 7.5 lower part. From then on, we will only eliminate variables.

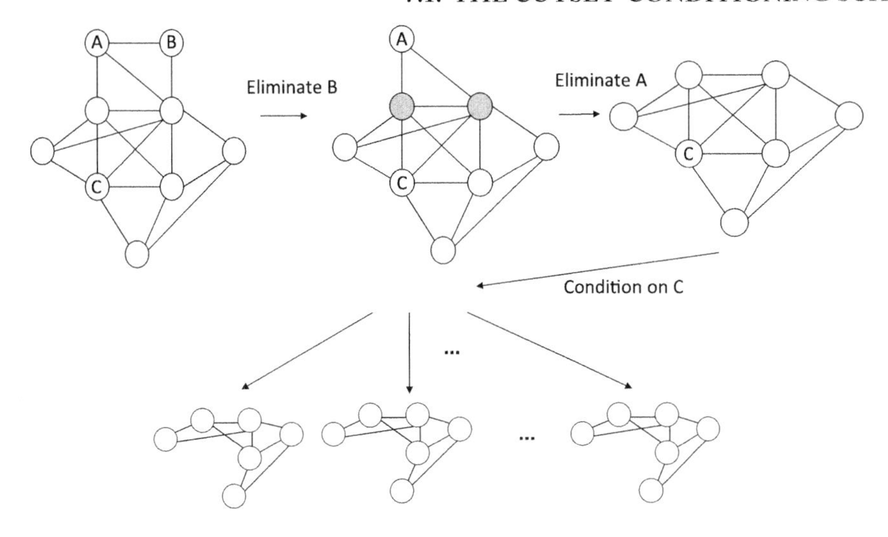

Figure 7.5: A graph (a), eliminating A (b), and conditioning on B (c).

It is possible to show that both *VEC(q)* and *ALT-VEC(q)* are basically the same performance-wise, if both use the same q-cutset. We refer the interested reader to [Mateescu and Dechter, 2005b].

In summary, the parameter q which bounds the conditioned induced-width can be used within VEC(q) to control the trade-off between search and inference. If d is the ordering used by $VEC(q)$ and if $q \geq w^*(d)$, the algorithm coincides with a pure inference algorithm such as bucket-elimination. As the control parameter q decreases, the algorithm requires less space and more time.

We can show that the size of the smallest cycle-cutset (1-cutset), c_1^* and the smallest induced width, w^* of a given graph, obey the inequality $1 + c_1^* \geq w^*$. More generally,

Theorem 7.7 *Given graph G, and denoting by c_q^* its minimal q-cutset then,*

$$1 + c_1^* \geq 2 + c_2^* \geq ...q + c_q^*, ... \geq w^* + c_{w*}^* = w^*.$$

Proof. Let's assume that we have a q-cutset of size c_q. Then if we remove it from the graph the result is a graph having a tree decomposition whose treewidth is bounded by q. Let's T be this decomposition where each cluter has size $q + 1$ or less. If we now take the q-cutset variables and add them back to every cluster of T, we will get a tree decomposition of the whole graph (exercise: show that) whose treewidth is $c_q + q$. Therefore, we showed that for *every* c_q-size q-cutset, there is a tree decomposition whose treewidth is $c_q + q$. In particular, for an optimal q-cutset of size c_q^*

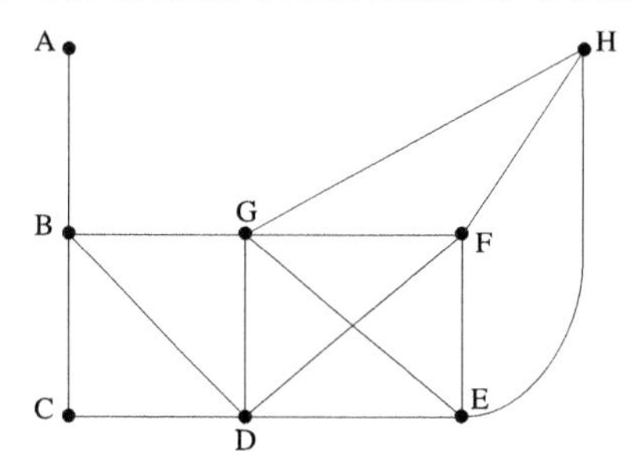

Figure 7.6: A primal constraint graph.

we have that $w*$, the treewidth obeys, $w* \leq c_q^* + q$. This does not complete the proof because we only showed that for every q, $w* \leq c_q^* + q$. But, if we remove even a single node from a minimal q-cutset whose size is c_q^*, we get a $q+1$ cutset by definition, whose size is $c_q^* - 1$. Therefore, $c_{q+1}^* \leq c_q^* - 1$. Adding q to both sides of the last inequality we get that for every $1 \leq q \leq w^*$, $q + c_q^* \geq q + 1 + c_{q+1}^*$, which completes the proof. □

The above relationship can suggest a point in which to trade space for time for given problem instances. In some extreme cases all the inequalities can become equalities. This is the case when we have a complete graph, for example. In a complete graph therefore it is better to use q=1 and resort to linear space search because the time complexity will not change if we apply full bucket-elimination, while the memory required would be heavy. For more on this see [Dechter and Fattah, 2001].

7.2 THE SUPER-CLUSTER SCHEMES

We now present an orthogonal approach for combining search and inference. In fact, the inference algorithm CTE in Figure 5.10 that processes a tree decomposition, already contains a hidden combination of variable elimination and search. It computes functions on the separators using variable elimination, and is therefore space exponential in the separator's size. The elimination of variables in any cluster can be carried out by search in time exponential in the cluster size but with a lower space complexity as presented in Theorem 5.28. Thus, one can trade even more space for time by allowing larger clusters, yet smaller separators.

Assume a problem whose tree decomposition has tree-width r and maximum separator size s. Assume further that our space restrictions do not allow the necessary $O(k^s)$ memory required when applying CTE on such a tree. One way to overcome this problem is to combine the nodes in the tree that are connected by large separators into a single cluster. The resulting tree decomposition has larger clusters but smaller separators.

We can get a sequence of tree decompositions parameterized by the sizes of their separators as follows. Let T be a tree decomposition of hypergraph $\mathcal{H}$. Let $s_0, s_1, ..., s_n$ be the sizes of the

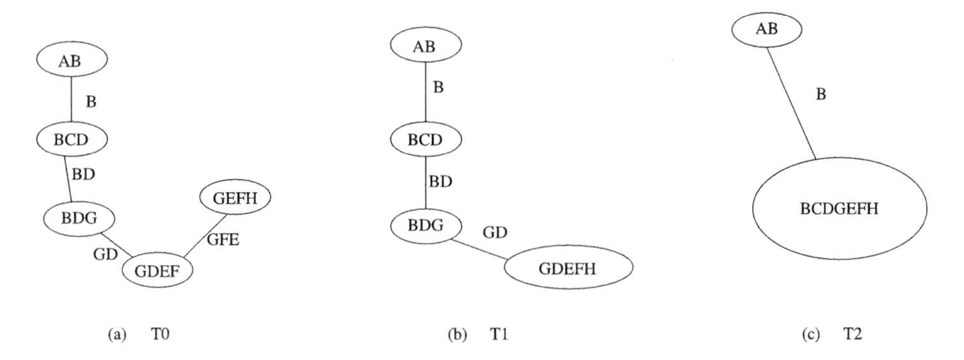

Figure 7.7: A tree decomposition with separators equal to (a) 3, (b) 2, and (c) 1.

separators in T, listed in strictly descending order. With each separator size s_i we associate a secondary tree decomposition T_i, generated by combining adjacent nodes whose separator sizes are strictly greater than s_i. We denote by r_i the largest set of variables in any cluster of T_i. Note that as s_i decreases r_i increases. Clearly, from Theorem 5.28 we have the following.

Theorem 7.8 *Given a tree decomposition T of graphical model having n variables and m functions, separator sizes $s_0, s_1, ..., s_t$ and secondary tree decompositions having a corresponding maximal cluster size, $r_0, r_1, ..., r_t$. The complexity of CTE when applied to each secondary tree decompositions T_i is $O(m \cdot deg \cdot exp(r_i))$ time, and $O(n \cdot exp(s_i))$ space (i ranges over all the secondary tree decomposition).*

We will call the resulting algorithm SUPER-CLUSTER TREE ELIMINATION(s), or $SCTE(s)$. It takes a primary tree decomposition and generates a tree decomposition whose separator's size is bounded by s. These cluster-trees can be processed by CTE. In the following example we assume that a naive depth-first search processes (enumerates the tuples in) each cluster when generating the messages over the separators.

Example 7.9 Consider the constraint problem having the constraint graph in Figure 7.6. The graph can be decomposed into the join-tree in Figure 7.7a. If we allow only separators of size 2, we get the join tree T_1 in Figure 7.7b. This structure suggests that applying CTE takes time exponential in the largest cluster, 5, while requiring space exponential in 2. If space considerations allow only singleton separators, we can use the secondary tree T_2 in Figure 7.7c. We conclude that the problem can be solved either in $O(k^4)$ time (k being the maximum domain size) and $O(k^3)$ space using T_0, or in $O(k^5)$ time and $O(k^2)$ space using T_1, or in $O(k^7)$ time and $O(k)$ space using T_2.

Algorithm $SCTE(s)$ (which is not presented explicitly) suggests the following new graph parameter.

Definition 7.10 Separator-bounded width. Given a graph G and a constant s, find a tree decomposition of G having the smallest induced-width, w_s^* whose separator size is bounded by s. The parameter w_s^* is the separator-bounded treewidth of the tree decomposition.

Finding a separator-bound treewidth w_s^* is hard, of course. This type of tree-decomposition is attractive because it requires only linear space. While we generally cannot find the best tree-decompositions having a bounded separators' size in polynomial time, this is possible for the extreme case when the separators are required to be singletons ($s = 1$), in which case the clusters are viewed as non-separable components [Even, 1979]. For the use of this special case see [Dechter, 2003].

7.3 TRADING TIME AND SPACE WITH AND/OR SEARCH

We will now generalize the above 3 algorithms (*VEC(q), ALT-VEC(q), SCTE(s)*) in several ways. First, whenever search is applied we replace standard OR search with AND/OR tree search. Second, when inference (e.g., bucket-elimination) is applied we can consider replacing it with the memory intensive AND/OR graph search. Finally, we describe all this having the more intensive counting queries, such as solution counting, probability of evidence or partition function, in mind.

Variable elimination and context-minimal AND/OR search. Before considering other variants of time-space trading algorithms we note that Variable Elimination (BE) and memory-intensive AND/OR Search (AO) can actually be viewed as searching the same AND/OR context minimal graph when they are guided by comparable variable-orderings when there is no determinism. Variable-elimination explores the search space bottom up in a breadth-first manner and AND/OR search explores it top-down in a depth-first manner (see [Mateescu, 2007]). Therefore, interchanging Bucket-elimination with context-based AO graph-search within a more global scheme can be entertained.

7.3.1 AND/OR CUTSET-CONDITIONING

The *VEC(q)* scheme we presented (either sequential or alternating) performs search on the cutset variables and exact inference on each of the conditioned subproblems. As we showed, if the *q-cutset* C_q is explored by linear space OR search, the total time complexity is $O(n \cdot k^{(|C_q|+q+1)})$, and the space complexity is $O(k^q)$. An immediate improvement to this scheme would be to enumerate the assignments to $\mathbf{C}_q$ by AND/OR search.

Example 7.11 Figure 7.8a shows two 3×3 grids, connected on the side node A. A cycle cutset must include at least two nodes from each grid, so the minimal cycle cutset contains three nodes:

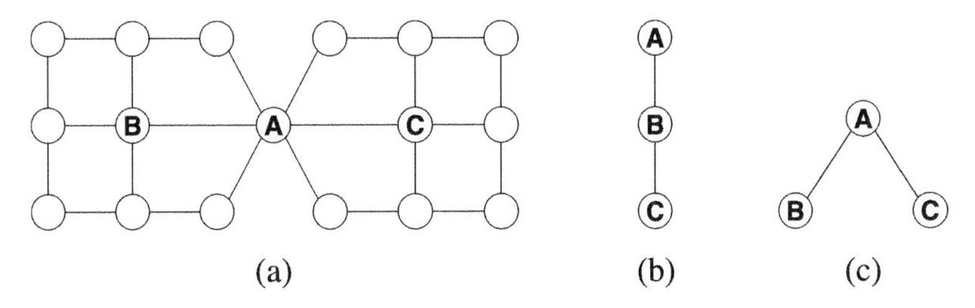

Figure 7.8: Traditional cycle cutset viewed as AND/OR tree.

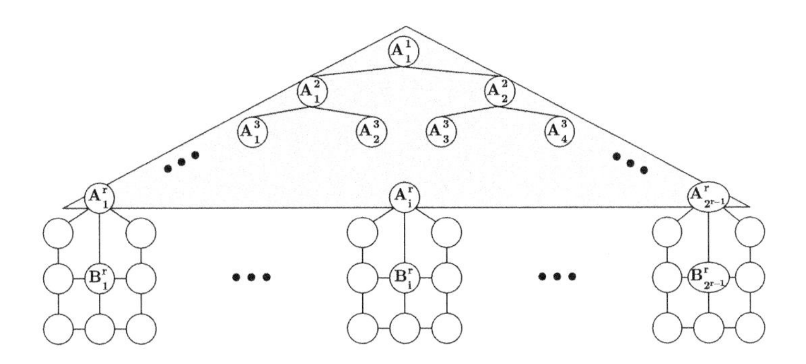

Figure 7.9: AND/OR cycle cutset.

the common node A and one more node from each grid, for example B and C. The cycle cutset scheme will enumerate all the assignments of $\{A, B, C\}$, as if these variables form the chain pseudo tree as in Figure 7.8b. However, if A is the first conditioning variable, the remaining sub-problem is split into two independent portions, so the cycle cutset $\{A, B, C\}$ can be organized as an AND/OR search space based on the pseudo tree in Figure 7.8c. If k is the maximum domain size of variables, the number of sub-problem displayed in Figure 7.8b is $O(k^3)$ while the number of sub-problems in Figure 7.8c is $O(k^2)$.

A more general and impressive is the example in Figure 7.11 where we have a complete binary tree of depth r. The leaf nodes root 3×3 grids. Since a cycle-cutset must contain two nodes from each grid an optimal cycle cutset would have $\mathcal{C} = \{A^r_1, \ldots, A^r_{2^{r-1}}, B^r_1, \ldots, B^r_{2^{r-1}}\}$, containing 2^r nodes, so the complexity of the VEC scheme is $O(k^{|\mathcal{C}|}) = O(k^{(2^r)})$. Consider now the AND/OR cycle cutset $AO\text{-}\mathcal{C} = \{A^j_i \mid j = 1, \ldots, r;\ i = 1, \ldots, 2^{j-1}\} \cup \{B^r_1, \ldots, B^r_{2^{r-1}}\}$, containing all the A and B nodes. A pseudo tree in this case is formed by the binary tree of A nodes, and the B nodes exactly in the same position as in the figure. The depth in this case is $r + 1$, so the complexity of exploring this as an AND/OR tree is $O(\exp(r + 1))$, even though the number of nodes is $|AO\text{-}\mathcal{C}| = |\mathcal{C}| + 2^{r-1} - 1$.

This suggests *AND/OR cutset* algorithm which can be far more effective than its OR version. The algorithm uses the notion of a start pseudo-tree.

Definition 7.12 Start pseudo-tree. Given an undirected graph $G = (\mathbf{X}, E)$, a directed rooted tree $T = (V, E')$, where $V \subseteq \mathbf{X}$, is called a *start pseudo-tree* if it has the same root and is a connected subgraph of some pseudo tree of G.

AND/OR cutset-conditioning schemes. The AND/OR cutset-conditioning schemes combines AND/OR search spaces with the cutset-conditioning idea. The conditioning (cutset) variables form a *start* pseudo tree. The remaining conditioned subproblems have bounded conditioned induced-width. Given a graphical model and a pseudo tree $\mathcal{T}$, we first find a start pseudo tree $\mathcal{T}_{start}$ such that the context of any node not in $\mathcal{T}_{start}$ contains at most q variables that are not in $\mathcal{T}_{start}$. This can be done by starting with the root of $\mathcal{T}$ and then including as many descendants as necessary in the start pseudo tree until the previous condition is met. $\mathcal{T}_{start}$ now forms a structured cutset, and when its variables are instantiated, the remaining conditioned subproblems has induced width bounded by q. The cutset variables can be explored by the linear space (no caching) AND/OR search, while the remaining variables outside the cutset, by using full caching, of size bounded by q. The cache tables need to be deleted and reallocated for each new conditioned subproblem (i.e., each new instantiation of the cutset variables).

We can explore the conditioned subproblems by bucket-elimination as before yielding an algorithm variant called (**AO-VEC(q)**). If instead we choose to explore the conditioned subproblem using the AND/OR context-minimal search, we call the algorithms by (**AO-CUTSET(q)**).

Theorem 7.13 Complexity. *Given a graphical model $\mathcal{M} = \langle \mathbf{X}, \mathbf{D}, \mathbf{F}, \bigotimes \rangle$ having primal graph G and given a q-cutset on a start pseudo-tree of height m, the time and space complexity of both* AO-VEC(q) *and* AO-CUTSET(q) *is $O(n \cdot k^{m+q+1})$ and $O(k^q)$, respectively.*

Finally, we can also augment the alternating variant (**ALT-VEC(q)**) so that its conditioning part will exploit an AND/OR search space. We will refer to this variant as *ALT-VEC-AO* and will elaborate on it shortly.

7.3.2 ALGORITHM ADAPTIVE CACHING ($AOC(q)$)

The cutset-consitioning principle and especially the variant AO-CUTSET(q), inspire a new algorithm, based on a more refined caching scheme for AND/OR search, which we call *Adaptive Caching* - $AOC(q)$. The algorithm integrates the idea of AND/OR cutset within AND/OR context minimal search. It caches some values even at nodes with contexts greater than the bound q as follows. Let's assume that $context(X) = [X_1 \dots X_k]$ and $k > q$. During AND/OR search, when variables $X_1, \dots, X_{k-q}$ are instantiated, they can be regarded as part of a cutset. The problem rooted by X_{k-q+1} can be solved in isolation, like a subproblem in the cutset scheme, after variables $X_1, \dots, X_{k-q}$ are assigned their current values in all the functions. In this subproblem,

```
Algorithm AOC(q).
    Input: M = ⟨X, D, F, ⊗⟩; G = (X, E); d = (X_1, ..., X_n); q

    Output: Updated belief for X_1
    Let T = T(G, d)            // create elimination tree for each X ∈ X do
      allocate a table for q-context(X)
    Initialize search with root of T;
    while search not finished do
        Pick next successor not yet visited                    // EXPAND;
1       Purge cache tables that are not valid;
2       if value in cache then
3           retrieve value; mark successors as visited;
4       while all successors visited do                        // PROPAGATE
5           Save value in cache;
6           Propagate value to parent;
```

$context(X) = [X_{k-q+1} \dots X_k]$, so it can be cached within space bounded by q. However, when the search retracts to X_{k-q} or above, the cache table for X needs to be deleted and will be re-allocated when a new subproblem rooted at X_{k-q+1} is solved. This is because the subproblem explored is conditioned on a different set of variables.

Definition 7.14 q-context, flag. Given a graphical model, a pseudo tree $\mathcal{T}$, a variable X and $context(X) = [X_1 \dots X_k]$, the *q-context* of X is:

$$q\text{-}context(X) = \begin{cases} [X_{k-q+1} \dots X_k], & if \quad q < k \\ q\text{-}context(X), & if \quad q \geq k \end{cases}.$$

X_{k-q} is called the **flag** of $q\text{-}context(X)$.

The high level pseudocode for $AOC(q)$ is given above. It is similar to AND/OR search based on full context. The difference is in the management of cache tables. Whenever a variable X is instantiated (when an AND node is reached), and for any variable Y such that X is the flag of $q\text{-}context(Y)$ the cache table is purged (reinitialized with a neutral value (line 1). Otherwise, the search proceeds as usual, retrieving values from cache if possible (line 3) or else continuing to expand, and propagating the values up when the search is completed for subproblem below (line 6).

Example 7.15 We will clarify here the distinction between AND/OR with full caching (AO), $AO - CUTSET(q)$ and Adaptive AND/OR Caching, AOC(q). We should note that the scope

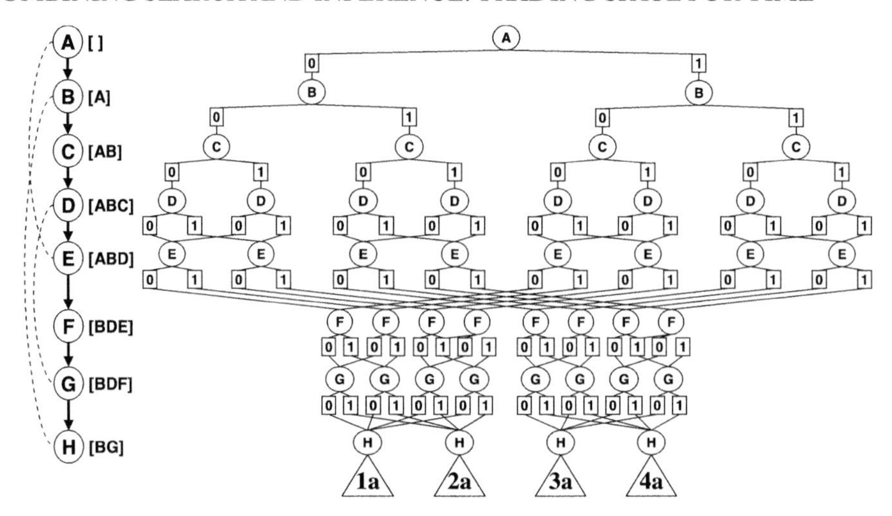

Figure 7.10: Context minimal graph (full caching).

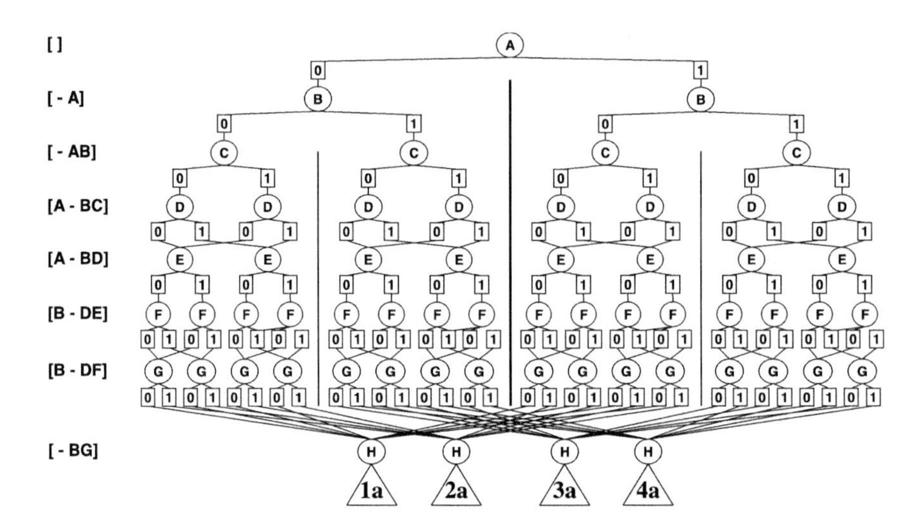

Figure 7.11:
AOC(2) graph (adaptive caching).

of a cache table is always a subset of the variables on the current path in the pseudo tree. Therefore, the caching method (e.g., full caching based on context (AO), cutset-conditioning cache, adaptive caching) is an orthogonal issue to that of the search space decomposition. Figure 7.10 shows a pseudo tree, with binary valued variables, the context for each variable, and the context minimal graph. If we assume the bound $q = 2$, some of the cache tables do not fit in memory. We could,

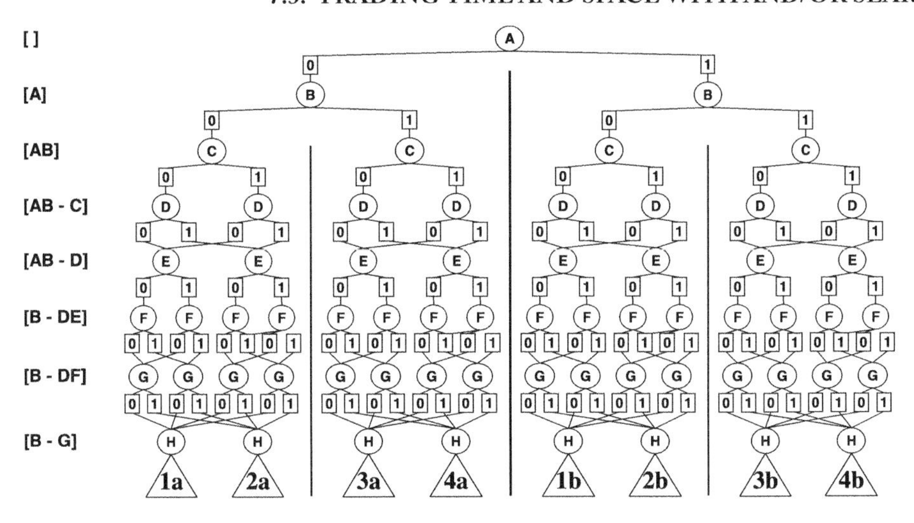

Figure 7.12: AOCutset(2) graph (AND/OR Cutset).

in this case, use **AO-CUTSET(2)** whose search-space is shown in Figure 7.12, that takes more time, but can be executed in the bounded memory. The cutset in this case is made of variables A and B, and we see four conditioned subproblems, the four columns, that are solved independently from one another (there is no sharing of subgraphs). Figure 7.11 shows the search space explored by $AOC(2)$, which falls between the previous two. It uses bounded memory, takes more time than full caching (as expected), but less time than $AO - CUTSET(2)$ (because the graph is smaller). This can be achieved because Adaptive Caching allows the sharing of subgraphs. Note that the cache table of H has the scope $[BG]$, which allows merging.

7.3.3 RELATIONS BETWEEN AOC(q), AO-ALT-VEC(q) AND AO-VEC(q)

We will now illustrate that there is no principled difference between some of the hybrid algorithms presented. Consider the graphical model given in Figure 7.13a having binary variables, the ordering $d_1 = (A, B, E, J, R, H, L, N, O, K, D, P, C, M, F, G)$, and the space limitation $q = 2$. A pseudo tree corresponding to this ordering is given in Figure 7.13b. The context of each node is shown in square brackets.

If we apply *AO-ALT-VEC(q)* along d_1 (eliminating from last to first), variables G, F and M can be eliminated. However, C cannot be eliminated, because it would produce a function with scope equal to its context, $[ABEHLKDP]$, violating the bound $q = 2$. AO-ALT-VEC switches to conditioning on C and all the functions that remain to be processed are modified accordingly, by instantiating C. The primal graph has two connected components now, shown in Figure 7.14. Notice that the pseudo trees are based on this new graph, and their shape changes from the original pseudo tree.

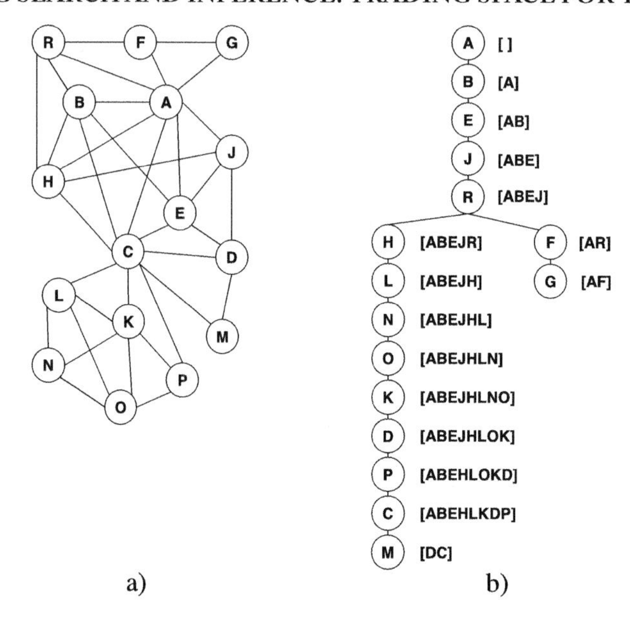

Figure 7.13: Primal graph and pseudo tree.

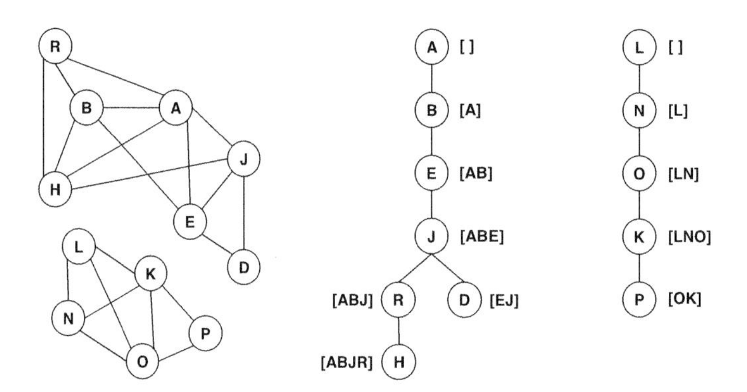

Figure 7.14: Components after conditioning on C.

Continuing with the ordering, P and D can be eliminated (one variable from each component), but then K cannot be eliminated. After conditioning on K, variables O, N and L can be eliminated (all from the same component), then H is conditioned (from the other component) and the rest of the variables are eliminated. To highlight the conditioning set, we will box its variables when writing the ordering, $d_1 = (A, B, E, J, R, \boxed{\mathrm{H}}, L, N, O, \boxed{\mathrm{K}}, D, P, \boxed{\mathrm{C}}, M, F, G)$.

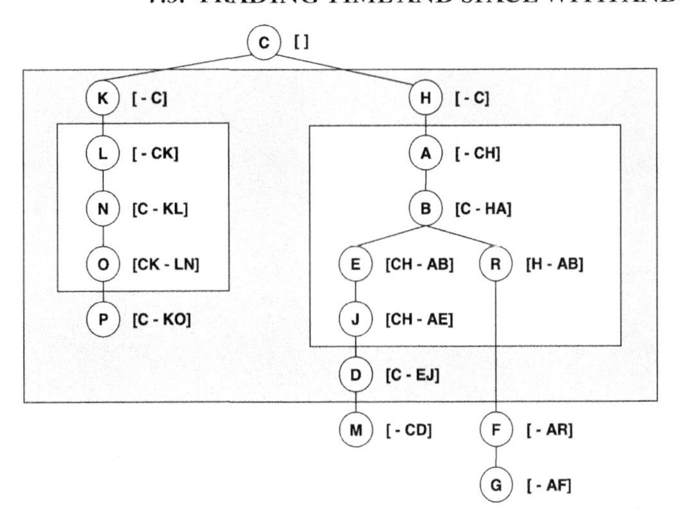

Figure 7.15: Pseudo tree for AOC(2).

If we take the conditioning set $[HKC]$ in the order imposed on it by d_1, reverse it and put it at the beginning of the ordering d_1, then we obtain:

$$d_2 = \left(\boxed{C}, \underline{\left[\boxed{K}, \underline{\left[\boxed{H}, \underline{[A,B,E,J,R]}_H, L,N,O \right]}_K, D,P \right]}_C, M,F,G \right),$$

where the indexed squared brackets together with the underlines represent subproblems that need to be solved multiple times, for each instantiation of the index variable.

Using ordering d_2 we will build a pseudo tree that can guide both $AO-VEC$ as well as $AOC(2)$, given in Figure 7.15. The outer box corresponds to the conditioning of C. The inner boxes correspond to conditioning on K and H, respectively. The context of each node is given in square brackets, and the *2-context* is on the right side of the dash. For example, $context(J) = [CH\text{-}AE]$, and 2-$context(J) = [AE]$. The context minimal graph corresponding to the execution of $AOC(2)$ is shown in Figure 7.16.

We can follow the execution of both $AOC(q)$ and $AO-VEC(q)$ along this context minimal graph. After conditioning on C, $AO-VEC(q)$ solves two subproblems (one for each value of C), which are the ones shown on the large rectangles.

If we change the ordering to $d_3 = (A, B, E, J, R, \boxed{H}, L, N, O, \boxed{K}, D, P, F, G, \boxed{C}, M)$, ($F$ and G are eliminated after conditioning on C), then the pseudo tree is the same as before, and the context minimal graph for AOC is still the one shown in Figure 7.16. Algorithm, $AO-$

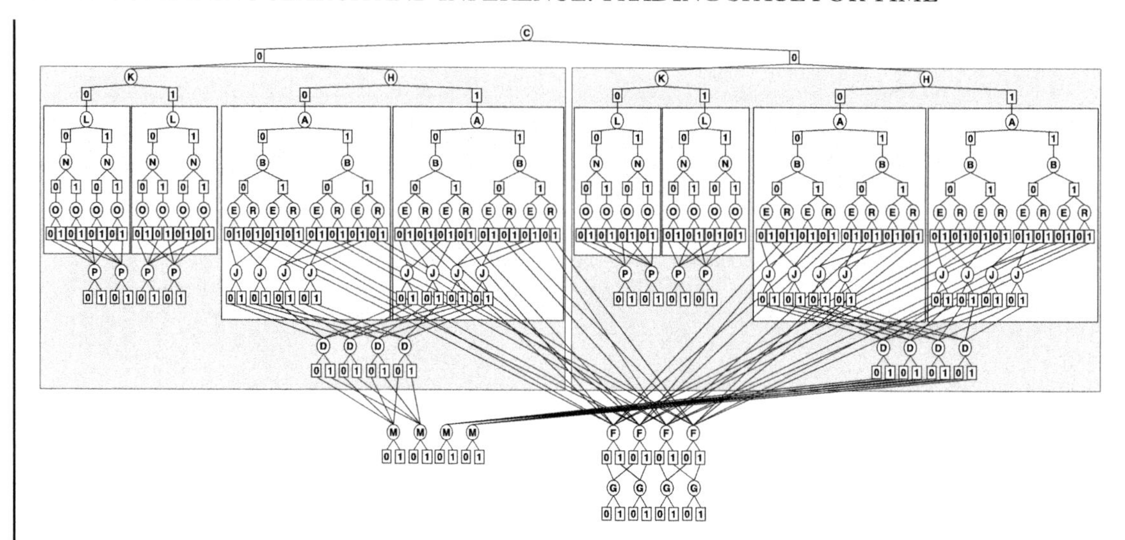

Figure 7.16: Context minimal graph.

$ALT - VEC(q)$ is given below (where $N_G(X_i)$ is the set of neighbors of X_i in the graph G). Note that the guiding pseudo tree is regenerated after each conditioning.

Algorithm $AO - ALT - VEC(q)$.

Input: $\mathcal{M} = \langle \mathbf{X}, \mathbf{D}, \mathbf{F} \rangle$; $G = (\mathbf{X}, E)$; $d = (X_1, \ldots, X_n)$; i

Output: Updated belief for X_1

Let $\mathcal{T} = \mathcal{T}(G, d)$ `// create elimination tree ;`

while $\mathcal{T}$ *not empty* **do**
  if $((\exists X_i$ *leaf in* $\mathcal{T}) \wedge (|N_G(X_i)| \leq i))$ **then** eliminate X_i **else** pick X_i leaf from $\mathcal{T}$;
  for *each* $x_i \in D_i$ **do**
    assign $X_i = x_i$;
    call $VEC(i)$ on each connected component of conditioned subproblem
  break;

Based on the previous example it is possible to show that

Theorem 7.16 ***AOC(q)* and *AO-VEC(q)* simulates *AO-ALT-VEC(q)*.** *Given a graphical model* $\mathcal{M} = \langle \mathbf{X}, \mathbf{D}, \mathbf{F}\ \bigotimes \rangle$ *with no determinism and an execution of* AO-VEC(q), *there exists a pseudo tree that guides an execution of* AOC(q) *that traverses the same context minimal graph. (For a proof see the appendix.)*

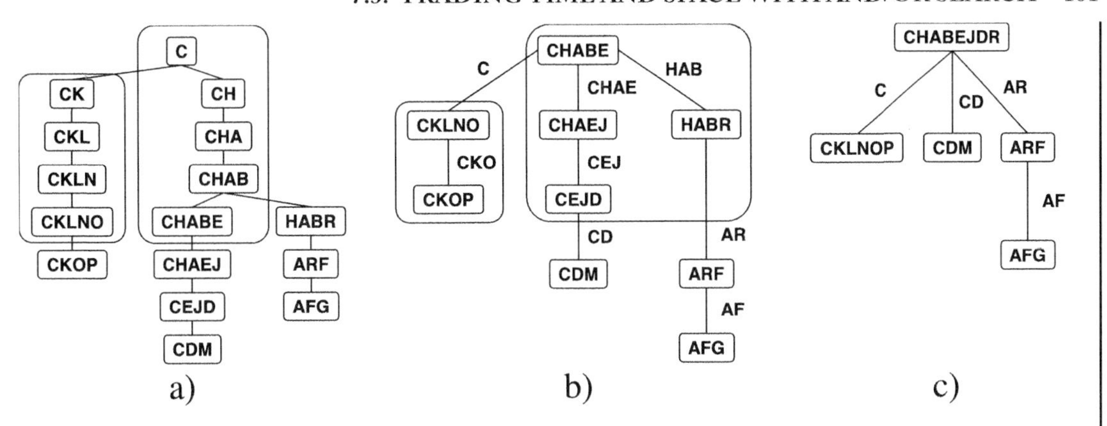

Figure 7.17: Tree decompositions: a) for d_2; b) maximal cliques only; c) secondary tree for $i = 2$.

7.3.4 AOC(q) COMPARED WITH STCE(q)

Algorithm SUPER-CTE(q), or $SCTE(q)$, described in Section 7.2. works on a tree-decomposition whose separator sizes are bounded by an integer q. The computation in each cluster was assumed initially to be naive, but it can be carried out by any search method. We can naively enumerate all the instantiations of the cluster, or, we can use more advanced method such as cutset-conditioning, or AND/OR search with adaptive caching. We refer to this resulting method as *SCTE with AO search*, or (SCTE-AO(q)). We next relate this variant to the AOC scheme.

Consider again at the example from Figures 7.15 and 7.16, and the ordering d_2. Because the induced parent set is equal to the context of a node, SCTE is equivalent to creating a cluster for each node in the pseudo tree from Figure 7.15, and labeling it with the variable and its context. The result is shown in Figure 7.17a. A better way to build a tree decomposition is to pick only the maximal cliques in the induced graph, and this is equivalent to collapsing neighboring subsumed clusters from Figure 7.17a, resulting in the tree decomposition in Figure 7.17b. If we want to run $STCE$ with bound $q = 2$, some of the separators are bigger than 2, so a secondary tree is obtained by merging clusters adjacent to large separators, obtaining the tree in Figure 7.17c. STCE(2) now runs by sending messages upwards, toward the root. Its execution, when augmented with AND/OR cutset in each cluster, can also be followed on the context minimal graph in Figure 7.16. The separators $[AF]$, $[AR]$ and $[CD]$ correspond to the contexts of G, F, and M in the graph in Figure 7.15. The root cluster $[CHABEJDR]$ corresponds to the part of the context minimal graph that contains all these variables. If this cluster would be processed by enumeration (OR search), it would result in a tree with $2^8 = 256$ leaves. However, when explored by AND/OR search with adaptive caching the context minimal graph of the cluster is much smaller, as can be seen in Figure 7.16. By comparing the underlying context minimal graphs, it can be shown that:

Theorem 7.17 *Given a graphical model $\mathcal{M} = \langle \mathbf{X}, \mathbf{D}, \mathbf{F} \rangle$ with no determinism, and an execution of $SCTE(q)$, there exists a pseudo tree that guides an execution of $AOC(q)$ and $AO - VEC(q)$ that* *traverses the same context minimal graph. (see [Mateescu and Dechter, 2007; Mateescu, 2007] for a proof).*

7.4 SUMMARY AND BIBLIOGRAPHICAL NOTES

In the past ten years, four types of algorithms have emerged, based on: (1) cycle-cutset and q-cutset [Dechter, 1990a; Pearl, 1988]; (2) alternating conditioning and elimination controlled by induced-width q [Fishelson and Geiger, 2002; Larrosa and Dechter, 2002; Rish and Dechter, 2000]; (3) recursive conditioning [Darwiche, 2001a], which was recently recast as context-based AND/OR search [Dechter and Mateescu, 2004]; and (4) varied separator-sets for tree decompositions [Dechter and Fattah, 2001]. This chapter is based on the above schemes and also, to a great extent on [Dechter and Mateescu, 2007b; Mateescu and Dechter, 2005a, 2007].

We presented 3 levels of algorithms trading space for time, based on combining search and inference. Algorithms VEC, ALT-VEC and SCTE, provided the principles integration schemes whose search component was naive. Algorithms *AO-VEC, AO-CUTSET* replaced the search component with AND/OR search and allowed the memory intensive component to be applied either via inference, yielding *AO-VEC,* or via context-based AND/OR search, yielding *AO-CUTSET*. The two final schemes of *AOC* and *SCTE-AO* provide an additional improvement.

We noted that there is no principled difference between memory-intensive search with fixed variable ordering and inference beyond: (1) different direction of exploring a common search space (top down for search vs. bottom-up for inference); (2) different assumption of control strategy (depth-first for search and breadth-first for inference). Also those differences occur only in the presence of determinism.

Adaptive Caching algorithm can be viewed as the most efficient AND/OR search algorithm, that exploits the available memory in the best way. Algorithm cutset-conditioning for Bayesian networks was introduced in [Pearl, 1988]. The cycle-cutset conditioning for constraint networks was introduced in [Dechter, 1990a]. Extensions to higher levels of cutset-conditioning appeared first in the context of satisfiability in [Rish and Dechter, 2000] and were subsequently addressed for constraint processing in [Larrosa and Dechter, 2003]. The cutset-conditioning scheme was used both for solving SAT problems and for optimization tasks [Larrosa and Dechter, 2001; Rish and Dechter, 2000] and is currently used for Bayesian networks applications [Fishelson and Geiger, 2002; Fishelson *et al.*, 2005]. Algorithms for finding small cycle-cutsets were proposed by [A. Becker and Geiger, 1999]. An algorithm for finding good w-cutset is given in [Bidyuk and Dechter, 2004].

7.5 APPENDIX: PROOFS

Proof of Theorem 7.16

The pseudo tree of $AOC(q)$ is obtained by reversing the conditioning set of $VEC(q)$ and placing it at the beginning of the ordering. The proof is by induction on the number of conditioning variables, by comparing the corresponding contexts of each variable.

Basis step. If there is no conditioning variable it was shown that AO is identical to VE is they are appleid along the same variable-ordering (thus the same pseudo-tree. If there is only one conditioning variable. Given the ordering $d = (X_1, \ldots, X_j, \ldots, X_n)$, let's say X_j is the conditioning variable.

(a) Consider $X \in \{X_{j+1}, \ldots, X_n\}$. The function recorded by $VEC(q)$ when eliminating X has the scope equal to the context of X in $AOC(q)$.

(b) For X_j, both $AO - VEC(q)$ and $AOC(q)$ will enumerate its domain, thus making the same effort.

(c) After X_j is instantiated by $VEC(q)$, the reduced subproblem (which may contain multiple connected components) can be solved by variable elimination alone. By the equivalence of AO to VE, variable elimination on this portion is identical to AND/OR search with full caching, which is exactly $AO - VEC(q)$ on the reduced subproblem.

From (a), (b) and (c), it follows that $AO - VEC(i)$ and $AOC(i)$ are identical if there is only one conditioning variable.

Inductive step. We assume that $VEC(q)$ and $AOC(q)$ are identical for any graphical model if there are at most k conditioning variables, and have to prove that the same is true for $k + 1$ conditioning variables.

If the ordering is $d = (X_1, \ldots, X_j, \ldots, X_n)$ and X_j is the last conditioning variable in the ordering, it follows (similar to the basis step) that $VEC(q)$ and $AOC(q)$ traverse the same search space with respect to variables $\{X_{j+1}, \ldots, X_n\}$, and also for X_j. The remaining conditioned subproblem now falls under the inductive hypothesis, which concludes the proof. Note that it is essential that $VEC(q)$ uses AND/OR over cutset, and is pseudo tree based, otherwise $AOC(q)$ is better. □

CHAPTER 8

Conclusion

We covered the principles of *exact* algorithms in graphical models, organized along the two styles of reasoning: *inference* and *search*. We focused on methods that are applicable to general graphical models, whose functions can come from a variety of frameworks and applications (constraints, Boolean, probabilities, costs etc.). These include, constraint networks and SAT models, Bayesian networks, Markov random fields, Cost networks, and Influence diagrams. Therefore, the primary features that capture structure in a unified way across all these models are graph features. The main graph property is the *induced-width* also known as *treewidth*, but we also showed the relevance of related features such as *height* of pseudo trees, *cycle-cutsets*, *q-cutsets* and *separator width*. We showed that both inference and search scheme are bounded exponentially by any of these parameters, and some combination of those hint at how we can trade memory for time.

With the exception of constraints, we did not discuss internal function structure as a potential feature. These function-structure features are sometimes addressed as *language* (e.g., Horn clauses, linear functions, convex functions) and can lead to various tractable classes. Other terms used are *context-sensitive* or *context specific independence*. In the constraint literature, tractability based on the language of constraints was investigated thoroughly (see Chapter 10 in[Dechter, 2003].) Likewise, focus on language is a central research activity in probabilistic reasoning. An example of a structure exploited in probabilistic graphical models are the sub-modular functions [Dughmi, 2009].

The next thing on our agenda is to extend the book with a second part focusing on approximation schemes. This obviously is necessary since exact algorithms cannot scale-up to many realistic applications that are complex and quite large and appropriately, current research centered on developing approximation schemes. But, we believe that **in order to have effective approximation algorithms we have to be equipped with the best exact algorithms, first.**

Approximation algorithms can be organized along the dimensions of inference and search as well. Given a general algorithmic architecture (such as Adaptive AND/OR search with caching (AOC(q)), or, alternatively, AO-VEC(q), we can approximate either the inference part or the search part or both, systematically yielding an ensemble of candidates approximation algorithms that can be studied. We can view messages-passing and variational algorithms such as generalized belief propagation, the mini-bucket and weighted mini-bucket schemes [Dechter and Rish, 2002; Liu and Ihler, 2011] as approximations that bound inference. We can view Monte Carlo sampling methods, as approximations to search. The hybrid schemes can be used to focus on approximating only those portions of the problem instance that appear non-tractable for exact processing. Namely, for a given problem instances, it can suggest a balance between approximate and exact and the type of approximation that should be utilized.

One should note that approximate reasoning in graphical modeling with any guarantees was shown to be hard as well [Dagum and Luby, 1993; Roth]. Yet, algorithms that generate bounds or anytime schemes that can improve their bounds if allowed more time, and even get to an exact solution when time permits, are highly desirable. Pointers to some literature on approximations can be found in recent PhD theses [Kask, 2001] [Bidyuk, 2006] and [Gogate, 2009] [Mateescu, 2007] [Marinescu, 2007] and in a variety of articles in the field such as (on message-passing variational approaches) [Mateescu *et al.*, 2010] [J. S. Yedidia and Weiss, 2005; M. J. Wainwright and Willskey, 2005; Wainwright and Jordan, 2008; Wainwright *et al.*, 2003], [Ihler *et al.*, 2012; Liu and Ihler, 2013], and [Sontag *et al.*, 2008]. On Sampling and hybrid of sampling and bounded inference see [Bidyuk and Dechter, 2007; Bidyuk *et al.*, 2010], [Gogate and Dechter, 2010, 2011, 2012]. On anytime schemes for optimization see [Marinescu and Dechter, 2009b; Otten and Dechter, 2012].

Bibliography

Bar-Yehuda R A. Becker and D. Geiger. Random algorithms for the loop-cutset problem. In *Uncertainty in AI (UAI'99)*, pages 81–89, 1999. DOI: 10.1613/jair.638. 146, 162

A. Darwiche. *Modeling and Reasoning with Bayesian Networks.* Cambridge University Press, 2009. DOI: 10.1017/CBO9780511811357. 7, 60, 133

S. M. Aji and R. J. McEliece. The generalized distributive law. *IEEE Transactions on Information Theory*, 46(2):325–343, 2000. DOI: 10.1109/18.825794. 28

D. Allen and A. Darwiche. New advances in inference by recursive conditioning. In *Proceedings of the 19th Conference on uncertainty in Artificial Intelligence (UAI03)*, pages 2–10, 2003. 138

S. A. Arnborg. Efficient algorithms for combinatorial problems on graphs with bounded decomposability - a survey. *BIT*, 25:2–23, 1985. DOI: 10.1007/BF01934985. 41, 43, 87, 102

R. Bar-Yehuda, D. Geiger, J. Naor, and R. M. Roth. Approximation algorithms for the feedback vertex set problem with applications to constraint satisfaction and bayesian inference. *SIAM J. Comput.*, 27(4):942–959, 1998. DOI: 10.1137/S0097539796305109. 145, 146

R. Bayardo and D. Miranker. A complexity analysis of space-bound learning algorithms for the constraint satisfaction problem. In *AAAI'96: Proceedings of the Thirteenth National Conference on Artificial Intelligence*, pages 298–304, 1996. 122, 140

A. Becker and D. Geiger. A sufficiently fast algorithm for finding close to optimal junction trees. In *Uncertainty in AI (UAI'96)*, pages 81–89, 1996. 41

A. Becker, R. Bar-Yehuda, and D. Geiger. Randomized algorithms for the loop cutset problem. *J. Artif. Intell. Res. (JAIR)*, 12:219–234, 2000. DOI: 10.1613/jair.638. 145, 146

C. Beeri, R. Fagin, D. Maier, and M. Yannakakis. On the desirability of acyclic database ochemes. *Journal of the ACM*, 30(3):479–513, 1983. DOI: 10.1145/2402.322389. 102

R.E. Bellman. *Dynamic Programming.* Princeton University Press, 1957. 65

E. Bensana, M. Lemaitre, and G. Verfaillie. Earth observation satellite management. *Constraints*, 4(3):293–299, 1999. DOI: 10.1023/A:1026488509554. 18, 124

U. Bertele and F. Brioschi. *Nonserial Dynamic Programming.* Academic Press, 1972. 46, 65, 102

B. Bidyuk and R. Dechter. On finding w-cutset in bayesian networks. In *Uncertainty in AI (UAI04)*, 2004. 146, 148, 162

B. Bidyuk and R. Dechter. Cutset sampling for bayesian networks. *J. Artif. Intell. Res. (JAIR)*, 28:1–48, 2007. DOI: 10.1613/jair.2149. 166

B. Bidyuk, R. Dechter, and E. Rollon. Active tuples-based scheme for bounding posterior beliefs. *J. Artif. Intell. Res. (JAIR)*, 39:335–371, 2010. DOI: 10.1613/jair.2945. 166

B. Bidyuk. Exploiting graph-cutsets for sampling-based approximations in bayesian networks. Technical report, PhD. thesis, Information and Computer Science, Universiy of California, Irvine, 2006. 166

S. Bistarelli, U. Montanari, and F. Rossi. Semiring-based constraint satisfaction and optimization. *Journal of the Association of Computing Machinery*, 44, No. 2:165–201, 1997. DOI: 10.1145/256303.256306. 11, 18, 28, 72

S. Bistarelli. *Semirings for Soft Constraint Solving and Programming (Lecture Notes in Computer Science*. Springer-Verlag, 2004. DOI: 10.1007/b95712. 28

H.L. Bodlaender. Treewidth: Algorithmic techniques and results. In *MFCS-97*, pages 19–36, 1997. DOI: 10.1007/BFb0029946. 102

C. Borgelt and R. Kruse. *Graphical Models: Methods for Data Analysis and Mining*. Wiley, April 2002.

C. Cannings, E.A. Thompson, and H.H. Skolnick. Probability functions on complex pedigrees. *Advances in Applied Probability*, 10:26–61, 1978. DOI: 10.2307/1426718. 72

M.-W. Chang, L.-A. Ratinov, and D. Roth. Structured learning with constrained conditional models. *Machine Learning*, 88(3):399–431, 2012. DOI: 10.1007/s10994-012-5296-5. 27

P. Beam H. Kautz Tian Sang, F. Bacchus and T. Piassi. Cobining component caching and clause learning for effective model counting. In *SAT 2004*, 2004. 140, 141

Z. Collin, R. Dechter, and S. Katz. On the feasibility of distributed constraint satisfaction. In *Proceedings of the twelfth International Conference of Artificial Intelligence (IJCAI-91)*, pages 318–324, Sidney, Australia, 1991. 140

Z. Collin, R. Dechter, and S. Katz. Self-stabilizing distributed constraint satisfaction. *The Chicago Journal of Theoretical Computer Science*, 3(4), special issue on self-stabilization, 1999. DOI: 10.4086/cjtcs.1999.010. 140

P. Dagum and M. Luby. Approximating probabilistic inference in bayesian belief networks is np-hard (research note). *Artificial Intelligence*, 60:141–153, 1993. DOI: 10.1016/0004-3702(93)90036-B. 166

A. Darwiche. Recursive conditioning. *Artificial Intelligence*, 125(1-2):5–41, 2001. DOI: 10.1016/S0004-3702(00)00069-2. 132, 140, 141, 162

M. Davis and H. Putnam. A computing procedure for quantification theory. *Journal of the Association of Computing Machinery*, 7(3), 1960. DOI: 10.1145/321033.321034. 46

S. de Givry, J. Larrosa, and T. Schiex. Solving max-sat as weighted csp. *In Principles and Practice of Constraint Programming (CP-2003)*, 2003. DOI: 10.1007/978-3-540-45193-8_25. 18

S. de Givry, I. Palhiere, Z. Vitezica, and T. Schiex. Mendelian error detection in complex pedigree using weighted constraint satisfaction techniques. In *ICLP Workshop on Constraint Based Methods for Bioinformatics*, 2005. DOI: 10.1007/978-3-540-45193-8_25. 18

R. Dechter and Y. El Fattah. Topological parameters for time-space tradeoff. *Artificial Intelligence*, pages 93–188, 2001. DOI: 10.1016/S0004-3702(00)00050-3. 150, 162

R. Dechter and R. Mateescu. The impact of and/or search spaces on constraint satisfaction and counting. In *Proceeding of Constraint Programming (CP2004)*, pages 731–736, 2004. DOI: 10.1007/978-3-540-30201-8_56. 162

R. Dechter and R. Mateescu. AND/OR search spaces for graphical models. *Artificial Intelligence*, 171(2-3):73–106, 2007. DOI: 10.1016/j.artint.2006.11.003. 127, 135, 140, 162

R. Dechter and J. Pearl. Network-based heuristics for constraint satisfaction problems. *Artificial Intelligence*, 34:1–38, 1987. DOI: 10.1016/0004-3702(87)90002-6. 29, 32, 46, 102, 135, 142

R. Dechter and J. Pearl. Tree clustering for constraint networks. *Artificial Intelligence*, pages 353–366, 1989. DOI: 10.1016/0004-3702(89)90037-4. 46, 102

R. Dechter and I. Rish. Directional resolution: The davis-putnam procedure, revisited. In *Principles of Knowledge Representation and Reasoning (KR-94)*, pages 134–145, 1994. 46

R. Dechter and I Rish. Mini-buckets: A general scheme for approximating inference. *Journal of the ACM*, pages 107–153, 2002. 165

R. Dechter and P. van Beek. Local and global relational consistency. *Theoretical Computer Science*, pages 283–308, 1997. DOI: 10.1016/S0304-3975(97)86737-0.

R. Dechter. Enhancement schemes for constraint processing: Backjumping, learning and cutset decomposition. *Artificial Intelligence*, 41:273–312, 1990. DOI: 10.1016/0004-3702(90)90046-3. 162

R. Dechter. Constraint networks. *Encyclopedia of Artificial Intelligence*, pages 276–285, 1992. DOI: 10.1002/9780470611821.fmatter. 140

R. Dechter. Bucket elimination: A unifying framework for probabilistic inference. In *Proc. Twelfth Conf. on Uncertainty in Artificial Intelligence*, pages 211–219, 1996. DOI: 10.1016/S0004-3702(99)00059-4. 28

R. Dechter. Bucket elimination: A unifying framework for reasoning. *Artificial Intelligence*, 113:41–85, 1999. DOI: 10.1016/S0004-3702(99)00059-4. 28, 71

R. Dechter. A new perspective on algorithms for optimizing policies under uncertainty. In *International Conference on Artificial Intelligence Planning Systems (AIPS-2000)*, pages 72–81, 2000. 6

R. Dechter. *Constraint Processing*. Morgan Kaufmann Publishers, 2003. 6, 7, 9, 15, 16, 18, 41, 46, 71, 98, 133, 134, 137, 140, 152, 165

R. Dechter. Tractable structures for constraint satisfaction problems. In *Handbook of Constraint Programming, part I, chapter 7*, pages 209–244. Elsevier, 2006. DOI: 10.1016/S1574-6526(06)80011-8.

S. Dughmi. Submodular functions: Extensions, distributions, and algorithms. a survey. *CoRR*, abs/0912.0322, 2009. 165

S. Even. Graph algorithms. In *Computer Science Press*, 1979. 152

S. Dalmo F. Bacchus and T. Piassi. Algorithms and complexity results for #sat and bayesian inference. In *FOCS 2003*, 2003. 141

S. Dalmo F. Bacchus and T. Piassi. Value elimination: Bayesian inference via backtracking search. In *Uncertainty in AI (UAI03)*, 2003. 141

M. Fishelson and D. Geiger. Exact genetic linkage computations for general pedigrees. *Bioinformatics*, 2002. DOI: 10.1093/bioinformatics/18.suppl_1.S189. 162

M. Fishelson and D. Geiger. Optimizing exact genetic linkage computations. *RECOMB*, pages 114–121, 2003. DOI: 10.1145/640075.640089. 148

M. Fishelson, N. Dovgolevsky, and D. Geiger. Maximum likelihood haplotyping for general pedigrees. *Human Heredity*, 2005. DOI: 10.1159/000084736. 162

E. C. Freuder and M. J. Quinn. The use of lineal spanning trees to represent constraint satisfaction problems. Technical Report 87-41, University of New Hampshire, Durham, 1987. 140

E. C. Freuder. A sufficient condition for backtrack-free search. *Journal of the ACM*, 29(1):24–32, 1982. DOI: 10.1145/322290.322292. 43

E. C. Freuder. A sufficient condition for backtrack-bounded search. *Journal of the ACM*, 32(1):755–761, 1985. DOI: 10.1145/4221.4225. 140

E. C. Freuder. Partial constraint satisfaction. *Artificial Intelligence*, 50:510–530, 1992. DOI: 10.1016/0004-3702(92)90004-H. 102

M. R Garey and D. S. Johnson. Computers and intractability: A guide to the theory of np-completeness. In *W. H. Freeman and Company, San Francisco*, 1979. 37, 144

N. Leone, G. Gottlob and F. Scarcello. A comparison of structural csp decomposition methods. *Artificial Intelligence*, pages 243–282, 2000. DOI: 10.1016/S0004-3702(00)00078-3. 102

V. Gogate and R. Dechter. On combining graph-based variance reduction schemes. *In 13th International Conference on Artificial Intelligence and Statistics (AISTATS)*, 9:257–264, 2010. 166

V. Gogate and R. Dechter. Samplesearch: Importance sampling in presence of determinism. *Artif. Intell.*, 175(2):694–729, 2011. DOI: 10.1016/j.artint.2010.10.009. 166

V. Gogate and R. Dechter. Importance sampling-based estimation over and/or search spaces for graphical models. *Artif. Intell.*, 184-185:38–77, 2012. DOI: 10.1016/j.artint.2012.03.001. 166

V. Gogate, R. Dechter, B. Bidyuk, C. Rindt, and J. Marca. Modeling transportation routines using hybrid dynamic mixed networks. In *UAI*, pages 217–224, 2005. 27

V. Gogate. Sampling algorithms for probabilistic graphical models with determinism. Technical report, PhD. thesis, Information and Computer Science, Universiy of California, Irvine, 2009. 166

H. Hasfsteinsson H.L. Bodlaender, J. R. Gilbert and T. Kloks. Approximating treewidth, pathwidth and minimum elimination tree-height. In *Technical report RUU-CS-91-1, Utrecht University*, 1991. DOI: 10.1006/jagm.1995.1009. 122

R. A. Howard and J. E. Matheson. *Influence diagrams*. 1984. 6, 9

A. Ihler, J. Hutchins, and P. Smyth. Learning to detect events with markov-modulated poisson processes. *ACM Trans. Knowl. Discov. Data*, 1(3):13, 2007. DOI: 10.1145/1297332.1297337.

A. T. Ihler, N. Flerova, R. Dechter, and L. Otten. Join-graph based cost-shifting schemes. In *UAI*, pages 397–406, 2012. 166

P. Meseguer J. Larrosa and M Sanchez. Pseudo-tree search with soft constraints. In *European conference on Artificial Intelligence (ECAI02)*, 2002. 140

W.T. Freeman J. S. Yedidia and Y. Weiss. Constructing free-energy approximations and generalized belief propagation algorithms. *IEEE Transaction on Information Theory*, pages 2282–2312, 2005. DOI: 10.1109/TIT.2005.850085. 166

F.V. Jensen. *Bayesian networks and decision graphs*. Springer-Verlag, New-York, 2001. 7

R. J. Bayardo Jr. and R. C. Schrag. Using csp look-back techniques to solve real world sat instances. In *14th National Conf. on Artificial Intelligence (AAAI97)*, pages 203–208, 1997. 135

J. Larrosa K. Kask, R. Dechter and A. Dechter. Unifying tree-decompositions for reasoning in graphical models. *Artificial Intelligence*, 166(1-2):165–193, 2005. DOI: 10.1016/j.artint.2005.04.004. 11, 28, 81, 93

H. Kamisetty, E. P Xing, and C. J. Langmead. Free energy estimates of all-atom protein structures using generalized belief propagation. In *Proceedings, Int'l Conf. on Res. in Comp. Mol. Bio.*, pages 366–380, 2007. DOI: 10.1007/978-3-540-71681-5_26.

K. Kask. Approximation algorithms for graphical models. Technical report, PhD thesis, Information and Computer Science, University of California, Irvine, California, 2001. 166

U. Kjæaerulff. Triangulation of graph-based algorithms giving small total state space. In *Technical Report 90-09, Department of Mathematics and computer Science, University of Aalborg, Denmark*, 1990. 43

D. Koller and N. Friedman. *Probabilistic Graphical Models*. MIT Press, 2009. 7, 60

J Larrosa and R. Dechter. Dynamic combination of search and variable-elimination in csp and max-csp. *Submitted*, 2001. 162

J. Larrosa and R. Dechter. Boosting search with variable-elimination. *Constraints*, 7(3-4):407–419, 2002. DOI: 10.1023/A:1020510611031. 162

J. Larrosa and R. Dechter. Boosting search with variable elimination in constraint optimization and constraint satisfaction problems. *Constraints*, 8(3):303–326, 2003. DOI: 10.1023/A:1025627211942. 162

J.-L. Lassez and M. Mahler. On fourier's algorithm for linear constraints. *Journal of Automated Reasoning*, 9, 1992. DOI: 10.1007/BF00245296. 41

S.L. Lauritzen and D.J. Spiegelhalter. Local computation with probabilities on graphical structures and their application to expert systems. *Journal of the Royal Statistical Society, Series B*, 50(2):157–224, 1988. 1, 102

Q. Liu and A. T. Ihler. Bounding the partition function using holder's inequality. In *ICML*, pages 849–856, 2011. 165

Q. Liu and A. T. Ihler. Variational algorithms for marginal map. *CoRR*, abs/1302.6584, 2013. 166

T. Jaakola M. J. Wainwright and A. S. Willskey. A new class of upper bounds on the log partition function. *IEEE Transactions on Information Theory*, pages 2313–2335, 2005. DOI: 10.1109/TIT.2005.850091. 166

D. Maier. The theory of relational databases. In *Computer Science Press, Rockville, MD*, 1983. 12, 84, 86, 102

R. Marinescu and R. Dechter. AND/OR branch-and-bound for graphical models. In *Proceedings of the Nineteenth International Joint Conference on Artificial Intelligence (IJCAI'05)*, pages 224–229, 2005. 129

R. Marinescu and R. Dechter. And/or branch-and-bound search for combinatorial optimization in graphical models. *Artif. Intell.*, 173(16-17):1457–1491, 2009. DOI: 10.1016/j.artint.2009.07.003. 129, 141

R. Marinescu and R. Dechter. Memory intensive and/or search for combinatorial optimization in graphical models. *Artif. Intell.*, 173(16-17):1492–1524, 2009. DOI: 10.1016/j.artint.2009.07.003. 129, 166

R. Marinescu. And/or search strategies for optimization in graphical models. Technical report, PhD. thesis, Information and Computer Science, Universiy of California, Irvine, 2007. 166

J. P. Marques-Silva and K. A. Sakalla. Grasp-a search algorithm for propositional satisfiability. *IEEE Transaction on Computers*, pages 506–521, 1999. DOI: 10.1109/12.769433. 135

R. Mateescu and R. Dechter. The relationship between AND/OR search and variable elimination. In *Proceedings of the Twenty First Conference on Uncertainty in Artificial Intelligence (UAI'05)*, pages 380–387, 2005. 162

R. Mateescu and R. Dechter. And/or cutset conditioning. In *International Joint Conference on Artificial Intelligence (Ijcai-2005)*, 2005. 149

R. Mateescu and R. Dechter. A comparison of time-space scheme for graphical models. In *Proceedings of the Twentieth International Joint Conference on Artificial Intelligence*, pages 2346–2352, 2007. 162

R. Mateescu, K. Kask, V. Gogate, and R. Dechter. Join-graph propagation algorithms. *J. Artif. Intell. Res. (JAIR)*, 37:279–328, 2010. DOI: 10.1613/jair.2842. 166

R. Mateescu. And/or search spaces for graphical models. Technical report, PhD. thesis, Information and Computer Science, Universiy of California, Irvine, 2007. DOI: 10.1016/j.artint.2006.11.003. 152, 162, 166

R. McEliece, D. Mackay, and J. Cheng. Turbo decoding as an instance of pearl's "belief propagation" algorithm. 16(2):140–152, February 1998.

L. G. Mitten. Composition principles for the synthesis of optimal multistage processes. *Operations Research*, 12:610–619, 1964. DOI: 10.1287/opre.12.4.610. 72

P. J. Modi, W. Shena, M. Tambea, and M. Yokoo. Adopt: asynchronous distributed constraint optimization with quality guarantees. *Artificial Intelligence*, 161:149–180, 2005. DOI: 10.1016/j.artint.2004.09.003. 140

U. Montanari. Networks of constraints: Fundamental properties and applications to picture processing. *Information Science*, 7(66):95–132, 1974. DOI: 10.1016/0020-0255(74)90008-5. 80, 98

K. P. Murphy. *Machine Learning; a probabilistic perspective*. 2012. 24

R.E. Neapolitan. *Learning Bayesian Networks*. Prentice hall series in Artificial Intelligence, 2000. 7

N. J. Nillson. *Principles of Artificial Intelligence*. Tioga, Palo Alto, CA, 1980. 108

L. Otten and R. Dechter. Anytime and/or depth-first search for combinatorial optimization. *AI Commun.*, 25(3):211–227, 2012. DOI: 10.3233/AIC-2012-0531. 129, 166

J. Pearl. *Probabilistic Reasoning in Intelligent Systems*. Morgan Kaufmann, 1988. 5, 6, 7, 9, 12, 20, 23, 24, 85, 98, 162

L. Portinale and A. Bobbio. Bayesian networks for dependency analysis: an application to digital control. In *Proceedings of the 15th Conference on Uncertainty in Artifi cial Intelligence (UAI99)*, pages 551–558, 1999. 27

A. Dechter R. Dechter and J. Pearl. Optimization in constraint networks. In *Influence Diagrams, Belief Nets and Decision Analysis*, pages 411–425. John Wiley & Sons, 1990. 72

B. D'Ambrosio R.D. Shachter and B.A. Del Favero. Symbolic probabilistic inference in belief networks. In *National Conference on Artificial Intelligence (AAAI'90)*, pages 126–131, 1990. 72

I. Rish and R. Dechter. Resolution vs. search; two strategies for sat. *Journal of Automated Reasoning*, 24(1/2):225–275, 2000. DOI: 10.1023/A:1006303512524. 37, 46, 135, 138, 162

D. Roth. On the hardness of approximate reasoning. *Artificial Intelligence*. DOI: 10.1016/0004-3702(94)00092-1. 166

D. G. Corneil S. A. Arnborg and A. Proskourowski. Complexity of finding embeddings in a k-tree. *SIAM Journal of Discrete Mathematics.*, 8:277–284, 1987. DOI: 10.1137/0608024. 43, 102

T. Sandholm. An algorithm for optimal winner determination in combinatorial auctions. *Proc. IJCAI-99*, pages 542–547, 1999. DOI: 10.1016/S0004-3702(01)00159-X. 18

L. K. Saul and M. I. Jordan. Learning in boltzmann trees. *Neural Computation*, 6:1173–1183, 1994. DOI: 10.1162/neco.1994.6.6.1174. 72

D. Scharstein and R. Szeliski. A taxonomy and evaluation of dense two-frame stereo correspondence algorithms. 47(1/2/3):7–42, April 2002.

R. Seidel. A new method for solving constraint satisfaction problems. In *International Joint Conference on Artificial Intelligece (Ijcai-81)*, pages 338–342, 1981. 46, 102

G. R. Shafer and P.P. Shenoy. Axioms for probability and belief-function propagation. volume 4, 1990. 28

P.P. Shenoy. Valuation-based systems for bayesian decision analysis. *Operations Research*, 40:463–484, 1992. DOI: 10.1287/opre.40.3.463. 10, 11, 28, 72

P.P. Shenoy. Binary join trees for computing marginals in the shenoy-shafer architecture. *International Journal of approximate reasoning*, pages 239–263, 1997. DOI: 10.1016/S0888-613X(97)89135-9. 93

K. Shoiket and D. Geiger. A proctical algorithm for finding optimal triangulations. In *Fourteenth National Conference on Artificial Intelligence (AAAI'97)*, pages 185–190, 1997. 41

J. Sivic, B. Russell, A. Efros, A. Zisserman, and W. Freeman. Discovering object categories in image collections. In *Proc. International Conference on Computer Vision*, 2005.

D. Sontag, T. Meltzer, A. Globerson, T. Jaakkola, and Y. Weiss. Tightening lp relaxations for map using message passing. In *UAI*, pages 503–510, 2008. 166

R. E. Tarjan and M. Yannakakis. Simple linear-time algorithms to test chordality of graphs, test acyclicity of hypergraphs and selectively reduce acyclic hypergraphs. *SIAM Journal of Computation.*, 13(3):566–579, 1984. DOI: 10.1137/0213035. 44, 45, 102

C. Terrioux and P. Jegou. Bounded backtracking for the valued constraint satsfaction problems. In *Constraint Progamming (CP2003)*, pages 709–723, 2003. 141

C. Terrioux and P. Jegou. Hybrid backtracking bounded by tree-decomposition of constraint networks. In *Artificial Intelligence*, 2003. DOI: 10.1016/S0004-3702(02)00400-9. 141

P. Thébault, S. de Givry, T. Schiex, and C. Gaspin. Combining constraint processing and pattern matching to describe and locate structured motifs in genomic sequences. In *Fifth IJCAI-05 Workshop on Modelling and Solving Problems with Constraints*, 2005. 18

P. Beam H. Kautz Tian Sang, F. Bacchus and T. Piassi. Cobining component caching and clause learning for effective model counting. In *SAT 2004*, 2004.

M. J. Wainwright and M. I. Jordan. Graphical models, exponential families, and variational inference. *Foundations and Trends in Machine Learning*, 1(1-2):1–305, 2008. DOI: 10.1561/2200000001. 166

M. J. Wainwright, T. Jaakkola, and A. S. Willsky. Tree-based reparameterization framework for analysis of sum-product and related algorithms. *IEEE Transactions on Information Theory*, 49(5):1120–1146, 2003. DOI: 10.1109/TIT.2003.810642. 166

Y. Weiss and J. Pearl. Belief propagation: technical perspective. *Commun. ACM*, 53(10):94, 2010. DOI: 10.1145/1831407.1831430. 98

A. Willsky. Multiresolution Markov models for signal and image processing. 90(8):1396–1458, August 2002.

C. Yanover and Y. Weiss. Approximate inference and protein folding. In *Proceedings of Neural Information Processing Systems Conference*, pages 84–86, 2002.

N.L. Zhang and D. Poole. Exploiting causal independence in bayesian network inference. *Journal of Artificial Intelligence Research (JAIR)*, 1996. DOI: 10.1613/jair.305. 72

Author's Biography

RINA DECHTER

Rina Dechter research centers on computational aspects of automated reasoning and knowledge representation including search, constraint processing, and probabilistic reasoning. She is a professor of computer science at the University of California, Irvine. She holds a Ph.D. from UCLA, an M.S. degree in applied mathematics from the Weizmann Institute, and a B.S. in mathematics and statistics from the Hebrew University in Jerusalem. She is an author of *Constraint Processing* published by Morgan Kaufmann (2003), has co-authored over 150 research papers, and has served on the editorial boards of: *Artificial Intelligence*, the *Constraint Journal*, *Journal of Artificial Intelligence Research (JAIR)*, and *Journal of Machine Learning Research (JMLR)*. She is a fellow of the American Association of Artificial Intelligence, was a Radcliffe Fellow 2005–2006, received the 2007 Association of Constraint Programming (ACP) Research Excellence Award, and she is a 2013 ACM Fellow. She has been Co-Editor-in-Chief of *Artificial Intelligence* since 2011. She is also co-editor with Hector Geffner and Joe Halpern of the book *Heuristics, Probability and Causality: A Tribute to Judea Pearl*, College Publications, 2010.

Printed in Great Britain
by Amazon